Anonymous

Tropical Nature

An Account of the most remarkable Phenomena of Life in the Western Tropics

Anonymous

Tropical Nature
An Account of the most remarkable Phenomena of Life in the Western Tropics

ISBN/EAN: 9783337216603

Printed in Europe, USA, Canada, Australia, Japan

Cover: Foto ©berggeist007 / pixelio.de

More available books at **www.hansebooks.com**

AN ACCOUNT

OF THE MOST REMARKABLE PHENOMENA OF LIFE

IN THE WESTERN TROPICS.

COMPILED FROM THE NARRATIVES
OF DISTINGUISHED TRAVELLERS AND OBSERVERS.

With Numerous Illustrations.

NEW YORK:
THOMAS NELSON AND SONS.
42 BLEECKER STREET.

CONTENTS.

CHAP.		PAGE
I.	LLANOS, SILVAS, PAMPAS, CAMPOS	1
II.	THE ANDES—THE MOUNTAINS OF BRAZIL	18
III.	RIVERS, GULFS, AND LAKES	34
IV.	CLIMATE, STORMS, WHIRLWINDS, WATERSPOUTS, MIRAGE, ZODIACAL LIGHT	55
V.	VEGETATION, FORESTS	78
VI.	THE INDIANS	103
VII.	QUADRUPEDS, MONKEYS	118
VIII.	BIRDS, REPTILES, INSECTS	147
IX.	VOLCANOES, EARTHQUAKES, METEORS, CAVERNS	175

ILLUSTRATIONS.

ALLIGATORS .	*Frontispiece.*
	PAGE
WILD HORSES AND CATTLE ON THE LLANOS	4
FELLING TREES IN A VIRGIN FOREST	6
THE CAMPO	9
VELLOZIA	11
OUSTITIS	13
RATTLESNAKE	14
BURNING OF WOODS ON THE HILLS NEAR THE CAMPO	14
FIRE ON THE CAMPO	16
DRY HAZE OVER THE SUN	17
CASCADE IN THE MOUNTAINS	22
STORM IN THE PARAMO	26
DISTANT VIEW OF THE ORGAN MOUNTAINS	29
CASCADE IN THE WOODS	32
STORM AT THE FOOT OF THE ORGAN MOUNTAINS	32
LIANAS	38
RAPIDS OF PIRAPORA	48
WAVES BREAKING ON THE SHORE	51
SUNSET AMONG THE MOUNTAINS	57
CLOUDS OVER THE MOUNTAINS	58
TWILIGHT RAYS	59
CLIFFS	60
HURRICANE IN THE ANTILLES	60
BOATS CAUGHT IN A STORM	62
ARBORESCENT LIGHTNING	64
WHIRLWIND	69
WHIRLWIND	70
WATERSPOUT ON LAND	71

ILLUSTRATIONS.

	PAGE
WATERSPOUT AT SEA	72
MIRAGE	74
THE SETTING MOON (LUMIÈRE CENDRÉE)	75
ZODIACAL LIGHT	76
THE MANGO AND THE JACK-TREE	80
VIRGIN FOREST IN BRAZIL	82
TROPICAL VEGETATION	85
LIANAS	87
TREE-FERN	90
MANGROVES	92
PALM-TREES ON THE SHORE	93
PALM-TREES	94
FOREST OF ARAUCARIAS	96
THE PAPAW TREE	98
BAMBOOS	100
INDIAN VILLAGE	105
INTERIOR OF INDIAN HUT	106
BOTOCUDOS SLEEPING	109
HUT OF SETTLED INDIANS	115
PUMA	119
PUMA WATCHING DEER DRINKING	121
JAGUAR WATCHING AN ENCAMPMENT	125
TAPIRS	127
TAMANDUA, OR ANT-BEAR	129
TATOU, OR ARMADILLO	139
CABIAI	140
COATIS, OR SOUTH AMERICAN RACOON	143
THE CONDOR	148
THE ALBATROSS	155
MARINE BIRDS	157
FLYING-FISH	158
HUNTING EMUS ON THE CAMPO	162
BOA CONSTRICTOR	172
VOLCANIC STORM	177
BOLIDE AT NINAS GERAËS	178
BOLIDE	181
CAVERN	182
STALACTITE CAVE NEAR JAGUERA	182
ENTRANCE OF CAVE	183

TROPICAL NATURE.

CHAPTER I.

LLANOS, SILVAS, PAMPAS, CAMPOS.

WHEREVER there is sufficient moisture to allow the productive powers of the soil to act, the regions bordering upon the Equator are characterised by luxuriance and variety; and when, as in some parts of America, the temperature is modified by the difference of elevation of various districts, those characteristics are enhanced to a degree of which the inhabitant of a temperate clime can scarcely form any commensurate idea. When Columbus first approached the coast of Paria and Cumana, in what has since been formed into the Republic of Columbia, he was so much delighted with the beauty and fertility of those regions that he believed he had discovered the garden in which Adam was originally placed by his Creator; and Brazil so abounds in all that can gratify the eye, the ear, and the palate, that Europeans who have settled there lose all wish to return to their native land; and if compelled by business to revisit it, are impatient to return, and 'celebrate Brazil as the fairest and most glorious country on the surface of the globe.'

In Central and South America all the features of nature are on a grand scale; the rivers are larger, the gulfs more capacious, the mountains higher and more precipitous, than almost all others in the world. The Amazon drains a surface of more than 2,000,000 of square miles, its course from its rise to its mouth is nearly 4000 miles, and the quantity of water it pours into the ocean is estimated to be ten times as much as is discharged by any other river. The mouth of the La Plata is nearly 170 miles in width, and the flow of its stream can be perceived at sea 200 miles from the coast. Thirty active volcanoes are situated amongst the mountains, and they contain precipitous rents of more than a mile in depth. The forests are some of the most extensive in the world, and abound in useful and ornamental woods, 117 different kinds having been collected, for the great Exhibition at Paris, from within the space of half a square mile; and 500 different species of palms—half the number of varieties known—being found in them. The different kinds of animals are more numerous than in any other part of the world, and while all the rivers of Europe do not furnish 150 species of fish, nearly 2000 are known to exist in the Amazon.

But notwithstanding the magnitude of the other features it presents, the most remarkable characteristic of South America is its plains. 'The high lands are a fringe of plateaux, and the gigantic valleys are low plains, broad, level tracts, that reach for thousands of miles in successive terraces, almost unbroken by hills.'

These level tracts are divided by the Andes and its offsets into five portions, each possessing peculiar characters of its own:—

1. The low country between the Andes and the coast, varying from 50 to 100 miles in width, and extending for about 4000 miles from north to south. This consists for the most part of a continuous waste, covered with a fine, light, yellow sand, on which no rain ever falls, and where no vegetation exists, except on the banks of the mountain torrents, many of which are dried up for a part of the year.

This sandy desert is intersected by hillocks or ridges, called Medanos, some of which are permanent, consisting of sand heaped up by the wind about a rocky nucleus; others are shifting, formed by the subsidence of columns of sand 80 to 100 feet in height, which are raised from time to time by the wind.

2. The Llanos of the Orinoco, extending from the mountains of Caracas to the left bank of the river. They were estimated by Humboldt to contain about 256,000 square miles, so level that often in a space of thirty leagues there is not a rising of a foot high; and the small eminences called Mesas—which separate the sources of rivers running into different oceans—are only a few feet in height, and are perceptible to strangers only on close observation.

The Llanos, unlike the Steppes of Asia, are strictly grassy plains, destitute of wood, or merely dotted here and there with palms, either of the Moriche or the Palma de Cobija, the wood of which is so hard that it is difficult to drive a nail into it. On approaching them from the north the astonished traveller finds that he has passed from a region abounding with organic life in rich luxuriance, to the borders of a treeless waste. Neither hill nor cliff rears its head above the boundless plain; only here and there, over a surface of more than 3000 square miles, appears any point sensibly higher than the rest. 'A llanero is happy only when, as expressed in the simple phraseology of the country, "he can see well around him." What appears to European eyes a covered country, slightly undulated by a few scattered hills, is to him a rugged region bristled with mountains. After having passed several months in the thick forests of the Orinoco, in places where one is accustomed, when at any distance from the river, to see the stars only in the zenith, as through the mouth of a well, a journey in the Llanos is peculiarly agreeable and attractive. The traveller experiences new sensations; and like the llanero, he enjoys the happiness of "seeing well around him."' But this enjoyment is not of long duration. 'When after eight or ten days' journey the traveller becomes accustomed to the mirage and the brilliant verdure of a few tufts of mauritia scattered from league to league he feels the want of more varied impressions ; he loves again to behold the great tropical trees, the wild rush of torrents, or hills and valleys cultivated by the hand of the labourer.'

Twice in every year the Llanos change their whole aspect, during one half appearing waste and barren like the Libyan desert ; during the other covered with verdure, like many of the elevated Steppes of Asia. 'When beneath the vertical rays of the tropical sun the parched sward crumbles into dust, the hardened soil cracks and bursts asunder as if rent by an earthquake. Gradually the pools of water, which had been protected from evaporation by the foliage of the fan palm, disappear. The crocodile and the boa lie wrapt in sleep, deeply buried in the dry earth. Shrouded in dark clouds of dust, and tortured by hunger and by burning thirst, oxen and horses scour the plain ; the one bellowing dismally, the other with outstretched necks snuffing the wind, endeavouring to detect the vicinity of some pool not wholly dried up.

'The mule, more cunning, adopts another means of allaying his thirst. There is a globular plant, the Melocactus, often 12 or 14 inches in diameter, which contains under its prickly coat an aqueous pulp. Carefully striking away the prickles with his fore-feet, he cautiously applies his lips to imbibe the cooling juice. He sometimes, however, pays dearly for the draught, and is often lamed by the cactus thorns.'

'Even when the heat of the day is succeeded by the freshness of the night, the wearied ox and horse have no repose. Huge bats now attack them and suck their blood ; or raise festering wounds, in which a host of stinging insects nestle and burrow. Such is the miserable state of these poor animals when the heat of the sun has dried up the moisture of the ground.'

Sometimes, towards the end of the dry season, extensive conflagrations take place, and dust-storms so dense that the lightning can only be heard, not seen; even when not a breath of air can be felt stirring, whirls of dust incessantly arise, increasing the effect of the suffocating heat.

When after a long drought the season of rain arrives, the scene suddenly changes. Scarcely is the earth moistened before it becomes covered with a variety of grasses—a 'sea of grass,' it is termed by the natives—affording rich pasture to the immense herds of horses, mules, and cattle, whose flesh and hides are chief articles of export of the towns on the coast.

WILD HORSES AND CATTLE ON THE LLANOS.

At times the clay on the banks of the morasses is seen to rise slowly in broad flakes. With a violent noise, like a small mud volcano, the upheaved earth is hurled into the air. Those who witness the phenomenon, and know the cause, fly from it, for a colossal water-snake, or a scaly crocodile, awakening from its summer trance, is about to burst forth.

Humboldt's host at Calabozo was sleeping with one of his friends on a bench, or couch covered with leather, when he 'was awakened early in the morning by a violent shaking and a horrible noise. Clods of earth were thrown into the middle of the hut. Presently a young crocodile, two or three feet long, issued from under the bed, darted at a dog which lay on the threshold of the door, and, missing him in the impetuosity of his spring, ran towards the beach to gain the river. On examining the spot where the *barbacoa*, or couch, was placed, the cause of this strange adventure was easily discovered. The ground was disturbed to a considerable depth. It was dried mud which had covered the crocodile in that state of lethargy or summer sleep, in which many of the species lie during the absence of the rains in the Llanos. The noise of men and horses, perhaps the smell of the dog, had aroused the crocodile. The hut being built at the edge of the pool, and inundated during part of the year, the crocodile had no doubt entered, at the time of the inundation, by the same opening at which it was seen to go out.'

When the rivers bounding the plain to the south, the Arauca, the Apure, and the Payara, overflow their banks, a part of the plain presents the appearance of an inland sea. The mares retreat with their foals to the higher banks. Day by day the dry surface diminishes in extent. The cattle, crowded together, swim for hours about the inundated plain, seeking a scanty nourishment from the flowering panicles of the grasses which rise above the waters. Many foals are drowned, many are seized by crocodiles and devoured. Horses and oxen may not unfrequently be seen, which have escaped from their attacks, bearing the marks of their teeth on their legs.

Agoutis; small-spotted antelopes; the shielded armadillo, which, rat-like, terrifies the hare in its subterranean retreat; herds of slothful chiguires (the Cavy capibarus); beautifully striped viverræ, whose pestilential odour infects the air; the great maneless lion (the Puma); the variegated jaguar (commonly known as the tiger), who hides in the luxuriant grass and darts with a cat-like bound on the passing prey, and whose strength enables him to drag to the summit of a hill the body of the young bull he has slain,—these and many other forms of animal life roam over the plain.

Families who live by raising cattle, and do not take part in agricultural pursuits, have congregated together in the middle of the Llanos, in small towns, which in the cultivated parts of Europe would scarcely be regarded as villages. Such are Calabozo, Villa del Pao, Saint Sebastian, and others. At distances of a day's journey from each other are found detached huts, covered with ox-hides. The cattle, horses, and mules, are not penned, but wander freely over an extent of several square leagues. There is nowhere any enclosure; men, naked to the waist and armed with a lance, ride over the savannahs to inspect the animals, to bring back any that wander too far away, and to brand with the owner's name all that are not already marked. These mulattos, who are known by the name of *peones llaneros*, are partly freedmen and partly slaves. They are constantly exposed to the burning heat of the tropical sun. Their food is meat dried in the air, and slightly salted; being always in the saddle, they fancy they cannot make the slightest exertion on foot. Their indolence is such that they do not dig wells, though they know that almost everywhere, at ten feet deep, fine springs are found. After suffering during one half of the year from the effect of inundations, they quietly resign themselves, during the other half, to the most distressing deprivation of water.

The numbers of the cattle in the Llanos are unknown; the large proprietors are quite ignorant of the numbers they possess. They only know that of the young cattle, which are branded every year with a mark peculiar to each herd. Some of them, in Humboldt's time, marked as many as 14,000 every year, and sold five or six thousand.

There is something awful, as well as sad and gloomy, in the aspect of the Llanos. Their uniformity, the total absence of all shrubs—in some places not even a solitary palm being visible; their stillness, scarcely a small cloud casting a shadow on the earth; the extremely small number of inhabitants; the fatigue of travelling beneath a burning sky, amidst an atmosphere darkened by dust; the boundless extent of the horizon, which seems for ever to fly before one; the lonely trunks of palm-trees, which seem all alike, and which it seems hopeless to attempt to reach, because they are confounded with others which rise on the horizon behind them: all these things combined make the Llanos appear even more extensive than they are in reality.

3. The basin of the Amazon, called the Silvas, containing about half a million of square miles, or six times the extent of France. 'This,' says Agassiz, 'is not a valley, in the ordinary sense, but a wide plain of rich alluvial soil, so level that the fall is little more than a foot in ten miles. The effect on the eye is, therefore, that of an absolute plain, and the flow of the water is so gentle that in many parts it is scarcely perceptible.'

The soil being rich, the climate moist, and the temperature high, every form of vegetation is rich beyond description. The forest is so dense as to be generally impenetrable. The largest trees bear brilliant scarlet, purple, rose-coloured, blue, and rich yellow flowers, blended with every conceivable shade of green. Palms and Melastomas, reeds 100 feet high, grasses 40 feet, and tree-ferns. 'The magnificent forest-trees are covered, and in some cases stifled, with overwhelming masses of

FELLING TREES IN A VIRGIN FOREST.

parasitical creepers; with Orchideæ, Araceæ, Tillandsias, epiphetic Cactuses, Piperonias, and Gesnerias, or with Lianas, including Bignonias, Passifloras, Aristolochias, &c. The Eriodendra saumauma puts forth no branches till it has overtopped every other tree of the vast Amazonian forests, over which it then predominates unrivalled. The Siphonia elastica, which yields caoutchouc, is indigenous in Brazil. That wonderful product, the Coca-leaf, or Ipadu, is obtained from the tropical valleys of the eastern slope of the Andes. The Yerba maté, or Paraguay tea, represents in its general use, as it partakes in the properties, the tea-plant of China.'

These forests abound with wild animals, amongst which are numerous monkeys of many varieties, but are thinly inhabited by wandering tribes of Indians.

4. The great southern plain, called the Pampas, watered by the Plata, extending from Buenos Ayres to the foot of the Andes, about 900 miles in breadth, and containing 1,217,000 square miles. These are for a great part of the space destitute of trees, and almost of water, with hardly a stone or a pebble to be seen for hundreds of miles. Sir Francis Head describes them as including three regions; the first of which, after leaving Buenos Ayres, 'is covered for 180 miles with clover and thistles; the second, for 450 miles, produces long grass; and the third, which reaches to the base of the Cordillera, is a grove of low trees and shrubs.' The trees and shrubs are evergreens, and the grass only changes from green to brown, from brown to green again. But the first region varies with the seasons in an extraordinary manner. 'In winter the leaves of the thistles are large and luxuriant, and the whole surface of the country has the rough appearance of a turnip-field. The clover in this season is extremely rich and strong; and the sight of the wild cattle grazing in full liberty on such pasture is very beautiful. In spring the clover has vanished, the leaves of the thistles have extended along the ground, and the country still looks like a rough crop of turnips. In less than a month the change is most extraordinary; the whole region becomes a luxuriant wood of enormous thistles, which have suddenly shot up to a height of ten or eleven feet, and are all in full bloom. The road or path is hemmed in on both sides; the view is completely obstructed; not an animal is to be seen; and the stems of the thistles are so close to each other, and so strong, that, independent of the prickles with which they are armed, they form an impenetrable barrier. The sudden growth of these plants is quite astonishing; and though it would be an unusual misfortune in military

history, yet it is really possible that an invading army, unacquainted with this country, might be imprisoned by these thistles before they had time to escape from them. The summer is not over before the scene undergoes another rapid change; the thistles suddenly lose their sap and verdure, their heads droop, the leaves shrink and fade, the stems become black and dead, and they remain rattling with the breeze one against another, until the violence of the pampero or hurricane levels them with the ground, where they rapidly decompose and disappear—the clover rushes up, and the scene is again verdant.'

'The vast region of grass in the Pampas for 450 miles is without a weed, and the region of wood is equally extraordinary. The trees are not crowded, but in their growth such beautiful order is observed that one may gallop between them in every direction. The young trees are rising up, others are flourishing in full vigour, and it is for some time that one looks in vain for those which in the great system of succession must necessarily somewhere or other be sinking towards decay. They are at last discovered, but their fate is not allowed to disfigure the general cheerfulness of the scene, and they are seen enjoying what may literally be termed a green old age. The extremities of their branches break off as they die, and when nothing is left but the hollow trunk it is still covered with twigs and leaves, and at last is gradually concealed from view by the young shoot, which, born under the shelter of its branches, now rises rapidly above it, and conceals its decay.'

Large herds of semi-wild horses and cattle graze on the Pampas, and a few wild animals, of which the biscachas are the most numerous. The ground is so much honeycombed with their burrows, that horsemen frequently get dangerous falls.

There is great difference between the temperature of the Pampas in summer and in winter. The winters are cold, and the ground is covered with frost in the morning. In the summer the heat is so oppressive that the horses and cattle seem exhausted by it.

The pampero is a violent south-west wind, which rushes over these plains with a force almost irresistible.

Fossil remains of extinct species abound. Mr. Darwin says, 'The whole area of the Pampas is one wide sepulchre of extinct gigantic quadrupeds.'

There are few inhabitants. The Gauchos, whose occupation is to look after the cattle and horses, are scattered over the northern parts, living a life of freedom and exertion, their food beef and water, their bed the ground.

In the south the Pampas Indians wander from place to place—a fierce and warlike people, feeding upon mares' flesh, living upon horseback, entirely naked, and armed with a spear eighteen feet long, which they manage with great dexterity.

5. The last, the largest, and the most interesting portion, is the high country of Brazil, to the east of the Parana and the Araguay, which forms about half of that immense empire. The average height is about

THE CAMPO.

2500 feet above the level of the sea, presenting alternate ridges and valleys, clothed with perpetual verdure. The declivities of the hills and the banks of the rivers are thickly covered with wood, opening into rich pasture-lands in the interior.

The aspect of these plains, or Campos, is very different from those of Caracas or Buenos Ayres. In some parts they are furrowed with small brooks bordered by rows of Mauritia Palms, with lofty trunks crowned by large leaves in the form of a fan; elsewhere large trees are scattered over them in the midst of the verdure like an English park; in others the ground is covered with low shrubs, intermingled with the Piqui, or Caryocar Brasiliensis, with its large palmated leaves of light green; so that the view from the small eminences which here and there occur is more pleasing than the unvarying prospect afforded by the treeless levels of the Llanos or the Pampas.

The animal kingdom, as well as the vegetable, also here presents great variety. Legions of birds of the most lively colours are always flying about amongst the trees and bushes: the emu, or American ostrich, and many species of gallinæ, run about the ground; while snakes of brilliant colours, but venomous bite or formidable size, twine through the grass and reeds, or bask in the sun. Deer and tapirs, jaguars, ant-eaters, and opossums, also inhabit this wilderness of beauty; while monkeys jump from tree to tree, and timid armadillos run about below.

All travellers represent the campos to be as healthful as they are beautiful. Though within the tropics the temperature is not oppressively hot, and the pure air restores to the exhausted traveller all his energies, mental and physical.

'In Rio,' says one traveller, 'I had heard much of the campos, so that I became almost impatient of the thick forests and narrow, swampy roads which we had travelled for about two leagues, when our guide made a short turn to the right, and suddenly the downs lay before us. We were at the foot of a short and steep hill. The morning was advanced and sultry, and among the woods not a breath of air was stirring. At once we were saluted by a fine bracing breeze in our faces, and hailed it with a burst of joy. We dismounted, and in the shade at the very verge of the forest refreshed ourselves and the horses. This was indeed a luxurious hour; I breathed ambrosial gales, and felt my nerves new strung. I had often heard of invalids who left the city in the last stages of debility, and on arriving at these salubrious regions were so much recovered that, according to their own phrase, they could buffet with and subdue a sturdy mule. Though in health, I experienced myself a wonderful renovation both of power and spirits.' 'The mornings and evenings,' says another, 'are the perfection of climate; the nights are cool, clear, and serene, as in the Arabian desert, without the sand.'

A remarkable contrast is presented by the campos to the traveller who approaches them from the dense virgin forests, into which the light of day penetrates only transiently and as through a veil. As he leaves the thick woods he enters upon a vast plain, with slight undulations, and bounded in the horizon by mountains to which the distance gives a bluish tinge. The higher parts are covered with a thick grassy carpet, which seems to indicate fertility; but it is mere appearance. The red clay bears only a shallow covering of productive soil, and the trees are unable to penetrate deeply, but throw out their roots along the surface in search of nourishment.

But in the lower grounds there are rich hollows, well fitted for cultivation, and on the slopes and the margins of the streams are clumps and borders of trees. Among these are often seen immense bombax-trees, covered with their large and brilliant flowers; and pao

VELLOZIA.

d'arcos, superb bignonias loaded with clusters of rose-coloured flowers at the ends of the branches, or other species with golden flowers. The elegant foliage of the mimosæ, the dark masses of myrtaceæ and terebinthaceæ, give these groups of trees a strange, and at the same time graceful aspect, in the eyes of a European. Many of the shrubs furnish useful products: catechu is obtained from the Acacia catechu, or khair-tree; the cantharides fly lives on a species of Vellozia. Some bear fruits which supply food for men or animals, some furnish gums, some dyes, some materials for soap, or for charcoal. One of the most valuable is the Arocira, of which the wood resists the weather, is very hard, and takes a fine polish; a decoction of the leaves serves to relieve rheumatic pains, and the gum acts as a preservative for ropes. Most of the shrubs and plants are medicinal—febrifuge, narcotic, or tonic; amongst them are those which furnish Cinchona and Ipecacuanha.

On passing into the open tracts the howling of herds of monkeys, the screaming of parrots, orioles, and toucans, the hammering of the woodpeckers, the deep metallic tones of the uraponga, or bell-bird, which resound through the forests, are no longer heard. The brilliant inhabitants of the plains are comparatively mute. Numerous humming-birds buzz like bees around the flowery shrubs; gay butterflies flutter from place to place; wasps fly in and out of their long nests suspended to the trees; large hornets hover over the ground, which is undermined with their cells. 'The red-capped and hooded fly-catcher, the barbudos (little sparrow-hawks), the rusty red or spotted cabori, bask on the shrubs during the heat of noon, and watch, concealed among the branches, for the small birds and insects which fly by; the tinamus walks slowly among the pine-apple plants; the cnapupés and emus in the grass; single toucans, seeking berries, hop among the branches; the purple tanagers follow each other in amorous pursuit from tree to tree; the caracari flying fearlessly about the roads, to settle upon the backs of the mules or oxen; small woodpeckers silently creep up the trees and look in the bark for insects; the rusty thrush, called Joaos de Barros, fixes its oven-shaped nest quite low between the branches; the siskin-like creeper slips imperceptibly from its nest (which, like that of the pigeons, is built of twigs, and hangs down from the branches to the length of several feet), to add a new division to it for this year; the caöha, sitting still on the tops of the trees, looks down after the serpents basking on the roads, on which, though poisonous, it feeds; and sometimes, when it sees people approaching, it sets up a cry of distress, like a human voice. It is only rarely that the tranquillity is disturbed, when garrulous orioles, little parrots and paroquets, arriving in flocks from the maize and cotton plantations in the neighbouring wood, alight upon the trees in the campo, and with terrible cries appear still to contend for the booty; or bands of restless hooded cuckoos, crowded together upon the branches, with noisy croaking defend their common nest, which is full of green-speckled eggs. Alarmed by the noise, or by passing travellers, numerous families of little pigeons, often no bigger than a sparrow, fly from bush to bush; the larger pigeons, seeking singly among the bushes for food, hasten in alarm to the tree-tops in the neighbouring wood, where their brilliant plumage shines in the sun; numerous flocks of little monkeys, searching for nuts or insects, run whistling and hissing to the recesses of the forest; the cavies hastily secrete themselves amongst the loose stones; the American ostriches, which herd in families, gallop at the slightest noise, like horses through the bushes, and over hills and valleys, accompanied by their young; the dicholopus, which pursues serpents, flies, sometimes sinking into the grass, sometimes rising into the trees, or rapidly climbing

OUSTITIS.

to the summits of the hills, where it sends forth its loud deceitful cry, resembling that of the bustard; the terrified armadillo runs fearfully about to seek a hiding-place, or, when the danger presses, shrinks into its armour; the tamandua, or ant-eater, runs heavily through the plain, and in case of need, lying on its back, threatens its pursuers with its long, sharp claws. Far from all noise, the slender deer, the black tapir, or the peccary, feed on the skirts of the forest. High above all this, the red-headed vulture soars aloft; hidden in the grass, the dangerous rattle-snake excites terror by its warning rattle; the gigantic boa sports suspended from the tree with its head upon the ground; and the crocodile, looking like the trunk of a tree, basks in the sun on the banks of the pools. After all this has during the day passed before the eyes of the traveller, the approach of night brings the chirping of the grasshoppers, the monotonous cry of the goat-sucker, the barking of the prowling wolf and of the shy fox, or the roaring of the ounce, to complete the picture of the animal kingdom in these peaceful plains.'— VON SPIX, vol. ii. 160–3.

At every step in this delightful wilderness, at every clump of trees, at

RATTLE-SNAKE.

every bush, new flowers, fresh aspects of beauty, are found; for variety is the marked character of tropical vegetation, and even the large masses of verdure have not the same uniformity of aspect which they generally, notwithstanding their own particular beauty, present in European countries. There is a grandeur in the vast expanse, together with beauty and variety in the details, which are not found combined in northern climates. Here the eye may rest upon the scene for hours without

BURNING OF WOODS ON THE HILLS NEAR THE CAMPO.

satiety, especially when viewing it from a rising ground, whilst the afternoon cloud casts a shifting shadow over the surface and intensifies the contrasts of colour, or when the changing tints of the 'after-glow' give warning of the approach of night.

Mr. Bates says that the changes of the seasons produce great changes in the appearance of the campos. The grass withers as the dry season advances, and the shrubs become parched and yellow. The trees, however, retain their verdure, and many of them flower at this period. The wet season begins suddenly about the beginning of February, generally commencing with violent squalls from the west, which in a few minutes drive all the boats on the river high up on the beach. The first storms are succeeded by a steady drizzle, after a week or two of which the appearance of the country is entirely changed; the grass springs up, with a great variety of quick-growing plants and creepers; the trees get a new coat of light green foliage; some of the shrubs begin to flower, and a great number of insects and birds are attracted. The foliage is most luxuriant in the months of June and July.

It is common in August or September, when the grass is dried up, to set fire to it, under the impression that the grass grows better afterwards; sometimes fires arise accidentally, or are caused merely for idle amusement. The flame then spreads till it is arrested by a brook, or a barren hill. It is a grand and impressive sight when it occurs at night. M. Liais mentions having witnessed one when a chain of mountains, near Marangaba, was covered with fire. The author of *Pictures from Cuba* thus describes one which he witnessed in that island:—'One evening, while watching the shadows of the trees and the tree-like vines in the lake, and the play of the dogs on the shore, I heard a rushing sound like the beating of many wings on the air, and looking in the direction whence it came, saw clouds of light-blue smoke rolling slowly up against the sky. In a few moments the southern sky was stained all over in black and gold with the thick smoke and leaping flames. We hurried to the house, and turning on the hill, saw a broad sheet of waving flame running all along the southern border of the lake, and reflected in the still water. More and more intense grew the conflagration, till it reddened the dark purple sky, and put out the stars above its path with its fiery glow. The graceful or fantastic shapes of the trees stood out finely from the wild back-ground, and from time to time a fresh gleam of flame, seen through the interstices of the thick low chapporal, would flash like the heart of a carbuncle.'

Many of the horses and cattle left in the campos are destroyed by these fires, notwithstanding they put forth all their power to endeavour to escape. Travellers who may happen to be in the neighbourhood are also exposed to great danger, as the conflagration spreads over the parched-up plains more quickly than the swiftest horse can gallop. Schomburg gives an account of a narrow escape which his party had on such an occasion, when they were saved only by one of the Indians dis-

FIRE ON THE CAMPO.

covering a small hill on which the vegetation was too scanty to afford fuel for the fire. Wild animals also perish, and the course of the fire is followed by voracious vultures, who pounce upon the snakes and lizards stifled by the blaze, and seem, when they dart upon their prey through clouds of smoke, as if they were voluntarily giving themselves to a fiery death.

These fires occasion dry fogs, by which the sky is completely veiled. Even in the middle of the day the sun appears pale and rayless, and may be steadily regarded with the naked eye. Its colour is reddish.

As it descends towards the horizon, its light is bedimmed so much, that its face can scarcely be perceived, but the edge is always clearly defined, and it often thus gradually disappears several degrees above the horizon.

Dry mists are not peculiar to America. They are very frequent in

DRY HAZE.

Africa, where M. d'Abbadie observed them in the eastern parts. They also occur in India, where there are also immense clouds of dust, especially in the plains neighbouring the Himalaya. According to the observations of M. Schlagentweit, whilst the sun appears red when seen through fogs, it looks blue through the clouds of dust, the cause of which diversity does not seem to be known with certainty.

CHAPTER II.

THE ANDES—THE MOUNTAINS OF BRAZIL.

SOUTH America, which is nearly double the extent of the whole of Europe, has nearly one-half of its surface covered with mountains, and amongst these are many which were long supposed to be the most lofty in the world, until observations made and measurements taken in the early part of the present century, showed that they are surpassed in height by some of the peaks in Asia. But though not the highest, they possess many features of great interest.

The loftiest of these mountains are in the range which runs nearly parallel to the west coast of the continent, known as the Andes, a name which is said to be derived from the Peruvian word *Anti*, the name of one of the races of the original inhabitants. This range may be considered to commence in Mexico, passing through the strip of land which connects North and South America, by the volcanic mountains of Guatemala. Entering South America at the Isthmus of Panama, it continues in an almost unbroken line to the southern extremity, a distance of 4500 miles.

The nearness of the Andes to the sea on the western front, and the consequent steepness of the ascent, gives them a magnificent aspect when approached from that side, which is thus described by Captain Basil Hall:—' On the 9th of June we sailed from Arica, and stood along shore to the north-west. In the evening of that day we had a fine view of the Cordillera, or highest ridge of the mountains, not less than between eighty and a hundred miles off. It was only when the ship was at a considerable distance from the shore that the higher Andes came in sight; for when near to it the lower ranges, themselves of great height, intercepted the remote view. But when we stretched off thirty or forty miles, these intermediate ridges sank into insignificance, while the chain of snowy peaks rose in great magnificence behind them. It sometimes even happened that the lower ranges, which had entirely obstructed the view of the Cordillera, when viewed at no great distance from the coast, were actually sunk below the horizon, by the curvature of the earth, when the distant ridges were still in sight, and more magnificent than ever. We were occasionally surprised, when we had little expectation of seeing the Cordillera, to behold the snowy tops towering above the clouds, and apparently so close that it required a considerable degree of experience, and a strong effort of reason, to remove them in imagination to their real distance. At first we were disappointed to find them so much lower than we had anticipated; but this arose from a misconception of their distance, and gave way gradually to the highest admiration when we became sensible by measurements, and by due reflection, how far they were from us . . .

'One morning we were much mortified when the day dawned to see no mountains in the eastern quarter, since we were not above a hundred miles from the shore; no land, however, could be distinguished. Presently the sun began to show himself above the horizon, and I have no language to tell the degree of interest which was excited when we discovered on his disc, as he rose, the outline of a distant summit of the Cordillera, clearly and sharply traced, but which was so far removed as to be totally invisible, except at this moment, when, being interposed between us and the sun, it intercepted a portion of his light, disclosed its situation for a few seconds, and then vanished again into thin air.'

When viewed from the eastern side the effect of their magnitude is diminished by the height of the country from which they rise. This is at such an altitude above the sea, that the elevation of the summits appears proportionally less. Chimborazo, Cotopaxi, and Antisana, though 6000 feet higher than Mont Blanc, and clothed like it with perpetual snow, appear to the traveller scarcely more sublime from the plains of Riobamba and Quito, than that mountain does from the vale of Chamouni. It requires some time for his imagination to realise their true grandeur.

In many parts the Andes are divided into two parallel chains of great elevation but small breadth, running about 100 miles apart from each other, and enclosing elevated plains of considerable extent. One of the most remarkable of these is the valley of Desarguardero, at an altitude of 13,000 feet: it extends for 500 miles with a breadth varying from 30 to 60 miles, the city of Potosi being situated on it at an elevation of 13,350 feet. A great part of this valley is occupied by the Lake Titicaca, the largest in South America, which covers a space of 3000 square miles, or about twenty times the size of the Lake of Geneva. This lake is in some parts 120 fathoms deep, and is an object of great veneration on the part of the natives, amongst whom there is a tradition that Manco Capac and his spouse made their first appearance in the country on an island in it, and gave forth thence their laws and institutions. Quito is situated in another of these plains, at an elevation of nearly 10,000 feet, extending for 200 miles in length and 30 in breadth, bounded by a series of the grandest volcanoes in the world. These parallel chains are found chiefly in the northern part of the country; for 2000 miles from Cape Horn to 20 degrees south, the chain is single.

Three minor chains branch off from the principal range nearly at right angles, by which, as already stated, the country to the east of the Andes is divided into three great plains or basins, i.e. the Llanos of Venezuela and Caracas, the Pampas of Buenos Ayres, and the Silvas, or woody basin of the Amazon.

In the neighbourhood of Caracas, one of the subsidiary ranges culminates in a double peak called the Silla (saddle), which was ascended by Humboldt and Bonpland during their visit. They could not find in the town a single person who had visited the summit, but having, through the assistance of the captain general, been furnished with some negroes as guides, they set out early in the morning of the 3rd of January. Their guides knew something of a path leading over the mountain near the western peak, but neither they, nor any of the militia, accustomed to the pursuit of smugglers over the mountain, had ever been on the eastern peak, which is the most elevated. The party consisted of eighteen persons, one of whom was a young Capuchin monk, who boasted much of the superior strength of European Spaniards over those born in America. He had provided himself with long strips of white paper, with which to leave marks on the way to indicate to those who might linger behind what direction they should follow. He had even promised to fire off some rockets, to announce to the whole town the success of the enterprise. But when they arrived at a steep ascent, covered with short grass, which afforded no good foothold, some of the party were discouraged and returned back, and the young monk stopped at a plantation, where he spent the rest of the day in watching through a glass the progress of those who persevered in the ascent.

They were much delayed by waiting for the loiterers, and when they resumed the ascent the weather was becoming cloudy, and mist was ascending from a small wood, in perpendicular streaks, as if fire had burst out at once in several different parts of it. Directing their course towards the eastern peak, with the cascade formed by the small river Chacaito on their left, they found only grassy savannahs up to a height of more than 6000 feet, when they saw in a ravine a grove of palm-trees. After proceeding across the savannahs for four hours the ascent became more gradual, and they entered a small wood in which they found a great number of beautiful plants—rhododendrons, thibaudias, andromedas, vacciniums, purple-flowered befarias, a heath-leaved hedyotis, eight feet high, and others. Amongst them is a shrub, ten or fifteen feet high, which the Creoles call *incienso* (incense): it is a species of Trixis, the flowers of which have the agreeable odour of storax. Whilst they were examining the plants in this wood, the sky became cloudy, and the thermometer fell to 50°. Quitting the little thicket, they found themselves again in a savannah. Passing into the hollow between the two peaks, they had great difficulty in making their way through the thick wood with which it was covered, consisting of a kind of plantain, 14 or 15 feet high. The negroes went before them to cut a way with their machetes. Whilst wandering in this wood they were suddenly enveloped in a thick mist, and were obliged to halt lest they should unawares find themselves on the edge of the precipice, which descends 6000 feet, almost perpendicularly, to the sea. They began to doubt whether they should reach the summit before night; but it being only two o'clock, and signs of an approaching change of weather appearing, they hoped to do so before sunset, and to return to the valley to pass the night. They therefore sent back part of the servants with instructions to meet them next morning with refreshments, but had scarcely done so when the east wind began to blow violently from the sea, the thermometer rose, and in less than two minutes the clouds were dispersed. They proceeded on their course, finding the vegetation less dense as they neared the summit, which they reached in about three quarters of an hour. Here the eye ranged over a vast extent, looking down on the sea to the north, and on the fertile valley of Caracas to the south. This mountain is surpassed in height by many others, but is remarkable for the enormous precipice on the side next the sea, so steep that when looking down on the houses below, it appears to be perpendicular. A rock of 1600 feet perpendicular height has been looked for in vain among the Alps, but this is between six and seven thousand feet. The horizon was invisible, being confounded with the sky, and so thick a fog soon arose that it would have been imprudent to remain on the edge of the precipice; they therefore commenced their descent about half-past four o'clock, but having delayed to botanise in the little wood, they were surprised by night at a height of five or six thousand feet, and were compelled to use great caution lest they should lose their footing and roll down some of the declivities. Some of the guides left them, lying down to sleep on the mountain, and those who remained with them, endeavouring to shorten the way, missed the path, and brought them down by a steep descent near the cascade of Chacaito, to which the darkness of night gave a grand and wild character, and they did not reach the foot of the mountain till ten o'clock.

In that part of the chain which is north of the Isthmus of Panama, it reaches its greatest height in the parallel of the city of Mexico, where some peaks are 17,000 feet above the sea, and many volcanoes are found. These also contain some of the richest mines that were known previous to the discoveries in California, the annual produce having been as much as six or seven millions sterling each of gold and of silver, besides iron, copper, and other metals.

CASCADE IN THE MOUNTAINS.

Coal has also been found at the height of nearly 15,000 feet, and fossils at between twelve and thirteen thousand feet.

A large population has been attracted by the mines in Mexico and Peru, and they present the extraordinary phenomenon of large communities of men subsisting in wealth and comfort at an elevation at which the inhabitants of plains find respiration difficult.

No glaciers are found in the Andes, except at the southern extremity. But in some parts there are awful perpendicular rents, called *quebrados*, forming narrow vales of immense depth, those at Chota and Cutaco being nearly a mile deep. These are toilsome and dangerous to cross; travellers generally pass them in chairs strapped on the backs of *cargueros*, a class of men who make it their occupation, and who clamber along these tremendous precipices with loads of 12, 14, or 18 stone. Hanging bridges of various materials are employed in some places. The fibres of the maguey-leaves are used for this purpose. These are first crushed between two stones, soaked in water till the surrounding matter easily separates from the fibres, when they are taken out, beaten with a stick, washed and dried; the ropes are then twisted by the hand, without the assistance of any machinery, additional fibres being added where necessary. A bridge of this kind near the village of Cochas was 38 yards across. On one side the five principal ropes, each about 12 inches in circumference, were fastened to a large beam laid on the ground, secured by two strong posts buried nearly to their tops; on the opposite side the beam was placed behind two small rocks. Across these five ropes were laid a number of the flower-stalks of the maguey, and upon them were strewed a quantity of old ropes and the fibrous parts of leaves, to preserve the stalks and the principal ropes. A network was placed on each side, to prevent passengers from falling into the river. Although the whole construction appears flimsy, the breadth being only five feet, droves of laden mules, as well as horned cattle, cross it. One of the largest of these bridges was between Lima and Cuzco, being 240 feet long and 9 feet broad. The bridges across the Mira are merely for foot-passengers, and are formed of the stems of the creeper called piquigua, which are remarkably fibrous and tough, and sometimes from 50 to 100 yards long. The stems are about half an inch in diameter, and are first beaten and then twisted, by which means a kind of cord is formed, five or six or more of which are combined to form a rope; two or more of these are laid side by side and covered with pieces of bamboo laid across, to form the pathway; hand-ropes made of piquigua being fastened to the side of the bridge to protect the passengers from falling over, which would otherwise be almost unavoidable, as they not only spring from the weight of the passenger, but from hanging loose they also swing from side to side. Some of them are formed like a ladder, and are crossed by stepping from one bar to another, a single hand-rope being the only safeguard against falling into the foaming stream, 80 or 100 feet below. Another mode of crossing the chasms or streams is called the *taravita*. This is formed by fixing the two ends of a rope of raw hide, or of piquigua, on opposite sides; on this is placed a pulley or a ring, to which is attached another rope passing through a pulley or ring on each side. To the pulley or ring on the large rope a basket is suspended, which is made of raw hide and called a *capache*. The passenger stands in this basket, and pulling the small rope he drags himself along, or else he is drawn across by persons on the other side: horses or cattle are placed in slings and suspended to the hook, and so drawn across.

At Icononzo, two days' journey from Bogota, a deep chasm, through which a small torrent rushes, is crossed by means of a natural bridge of stone, 312 feet above the surface of the water. This arch, which is forty-six feet in length and nearly forty in breadth, with a thickness of about seven feet, appears to be formed by a stratum of compact stone which has resisted the force by which the chasm was made. Sixty-feet below the arch is another, consisting of three enormous masses of rock, which have fallen in such a manner as to support each other. In the middle is a hollow about eight yards square, affording a view of the torrent below, apparently flowing through a dark cavern, whence arises a mournful noise, caused by the flight of numberless birds flying over the surface of the water, at such a depth that they can be seen only by throwing down rockets to illuminate the sides of the crevice.

A few leagues from Bogota, a small river, formed by the water collected in the valley, forces its way through the surrounding mountains by a narrow outlet, and forms the celebrated fall of Tequendama. The river, which at a short distance is 140 feet in width, is contracted into a narrow and deep bed only 40 feet broad, and precipitates itself down a perpendicular rock to the depth of nearly 600 feet at two bounds. 'This vast body of water at first forms a broad arch of a glassy appearance; a little lower down it assumes a fleecy form; and ultimately, in its progress downwards, shoots forth into millions of tubular shapes, which chase each other more like sky-rockets than anything else. The changes are as singularly beautiful as they are varied, owing to the difference of gravitation and the rapid evaporation which takes place before reaching the bottom. The noise with which this immense body of water falls is quite astounding; sending up dense clouds of vapour, which rise to a considerable height and mingle with the atmosphere, forming in their ascent the most beautiful rainbows. It is asserted that experiments have more than once been made of forcing a bullock into the stream, and that no vestige of him has been found at the bottom but a few of his bones. From the rocky sides of the immense basin hung with shrubs and bushes, numerous springs and tributary streams add to the grand effect. At the bottom the water which runs off rushes impetuously along a stony bed, overhung with trees, and loses itself in a dark winding of the rock. From the level of the river the mountains rise to a great height, and are completely covered with wood: and at one opening is an extensive prospect, which, on a clear day, includes some distant mountains in the province of Antioquia, whose summits are clothed with perpetual snow. Hovering over the frightful chasm are various birds of the most beautiful plumage, peculiar to the spot. A few feeble rays of noon fall on the bottom of the crevice. The solitude of the place, the richness of the vegetation, and the dreadful roar that strikes upon the ear, contribute to render the foot of the cataract of Tequendama one of the wildest scenes that can be found in the Cordilleras.'

Little or no rain falls on the western side of the Peruvian Andes, and the rivers are short, shallow, and rapid; consequently the vegetation is scanty and almost entirely confined to the neighbourhood of the streams. But on the elevated plains between the transverse groups sudden and violent changes of weather take place; clouds of black vapour are carried by fierce winds over the plains; snow, rain, hail, sleet, gleams of sunshine, rapidly succeed one another in the course of a few hours; thunderstorms loud and awful, come on without warning, and notwithstanding the thinness of the air, the crash of the peals is appalling, the lightning runs along the scorched grass, and sometimes issuing from the ground destroys a team of mules or a flock of sheep at one flash. Thick fogs frequently occur, lasting for several days, and terrific hail-storms; the hailstones being not only of different forms, generally much flattened by rotation, but also run together into thin plates of ice, which cut the face and hands in their fall. Mr. Darwin relates that one of his guides, when a boy of fourteen years of age, was passing the Cordillera with a party in the month of May, when a furious wind arose, so that the men could hardly cling on their mules, and stones were flying along the ground. The day was cloudless, and not a speck of snow fell, but the temperature was low. The gale lasted more than a day; the men began to lose their strength, and the mules would not move onwards. His brother tried to return, but he perished, and his body was found two years afterwards, lying by the side of his mule near the road, with the bridle still in his hand. Two other men in the party lost their fingers and toes; and out of two hundred mules and thirty cows, only fourteen mules escaped alive. Many years before the whole of a large party were

STORM IN THE PARAMO.

supposed to have perished in a similar manner, but their bodies were never discovered.

The variety of forms of vegetation which the Andes present to view, when ascending from height to height, is unsurpassed even in the Himalayas. 'A person who for the first time climbs the mountains of Switzerland is astonished to witness, in the space perhaps of a few hours, so rapid a change of climate, and such a wide range of vegetable productions. He may begin his ascent from the midst of warm vineyards, and pass through a succession of chestnuts, oaks, and beeches, till he gains the elevation of the hardy pines and stunted birches, or treads on Alpine pastures, extending to the borders of perpetual snow.

But within the tropics everything is formed on a grander scale. The boundary of the permanent congelation is 7500 feet higher at the Equator than at a mean latitude of 45°. Under a burning sun ananas and plantains grow profusely near the shore; oranges and limes occur a little higher; then succeed fields of maize and luxuriant wheat; and the traveller has actually reached the high plain of Mexico, or the still loftier vale of Quito, before he finds a climate analogous to that of Bourdeaux or Geneva. Now only commences the series of plants which inhabit the central parts of Europe.' Out of 327 genera of plants found at the height of 7800 feet and upwards, 180 are common to the temperate zone.

'The finest quality of cascarilla is found only on the eastern side of the Andes; of the indigo there is no end: I can say the same of the cotton and the rice. The precious balsam of copaiba, the sarsaparilla, the gum elastic, and the most fragrant species of vanilla, are all produced in an extraordinary abundance in these regions. The mighty forests which line the shores of the rivers abound in the finest timber for all uses, especially for ship-building, and in trees distilling the most aromatic and medicinal gums. Among others there is a species of cinnamon, called by the natives *canela de clavo*, which only differs in the thickness of the bark and its darker colour, according to its age, from that found in the East Indies, and which is as fragrant as the spice (clove) from which it takes its name.'—(THEO. HEINKE, in *Journal of Geographical Society*.) The coca leaf is also obtained from a plant which grows in this part. It is a bush about 6 or 8 feet high, somewhat like the blackthorn. The leaves are carefully gathered and dried, and packed in woollen sacks and covered with sand. Every Indian carries a leathern pouch with a supply of leaves, and a small flask gourd filled with unslaked lime. Three or four times a-day they rest to chew it. Some attribute dreadful effects to this habit; others say that if used in moderation it is beneficial. Dr. Smith during several years' residence in the district never saw an instance in which either mania or tremor was produced by it; and Von Tschudi, during his travels, was accustomed to use an infusion of the leaves to counteract the effect of the rarified air of the mountains, and found no ill effects to follow.

The mountains to the east of the Andes are of comparatively small extent and elevation. The land rises gradually from the east coast, and at no great distance attains a height of five or six thousand feet, forming a chain, called the Serro do Mar, nearly parallel with the coast, and extending southwards nearly to the Rio de la Plata. A little further inland is the chain of Villa Rica, running almost parallel with the coast chain, between which and the eastern foot of the Andes the country forms a vast plateau, of an average height of more than 3000 feet above the sea, covered with forests of magnificent trees, bound together by tangled creeping and parasitical plants. This is intersected by a few ridges of limited extent, but with no overtowering peaks like those in the Cordilleras. The difference of their elevation is comparatively inconsiderable; the average height of the five groups east of the Andes being from 3000 to 4200 feet, and the highest points between 6000 and 8000 feet above the sea; none of them entering within the line of perpetual snow.

On entering the Bay of Rio de Janeiro the Organ mountains are seen at about 18 leagues' distance. The description of them given by M. Liais will serve to convey an idea of the character of the mountainous districts of Brazil. These mountains appear as if arising directly from the sea, and seen from a distance their aspect is very imposing. Especially when they are observed from an elevated point commanding the numerous small islands in the bay, the combination of the distant blue mountains, and the green islands rising from the calm water, forms a striking view, and gives a strong impression of the greatness of the convulsion which has produced them.

A steamboat plies from Rio to the little harbour of Piedade, whence a diligence conveys travellers to the foot of the mountains, passing over level ground to the little town of Magé, at about two leagues' distance. At the sides of the road are seen groups of rhexia, with bunches of large rose-coloured and violet flowers; of Clerodendrum fragrans, covered with large sprays of double white flowers, smelling like orange; of Datura arborea, with fragrant white bells; innumerable brilliant yellow flowers of Thunbergia alata enamelling the ground. These three plants grow with such profusion in the neighbourhood of Rio that, if we did not know their native country, it would be difficult to believe that they are not indigenous. After having passed the town the road gently ascends for about two leagues to the port of Barrieres, at the foot of the mountains.

'Let us imagine on the right,' says M. Liais, 'an immense cone of granite, with bare sides, rising from a narrow base to more than 1200 mètres in height; behind it, a spur of the chain with which it is connected for a portion of its height, the ground of which is covered with a dense forest; then above, and on the same side, the summits of two

DISTANT VIEW OF THE ORGAN MOUNTAINS.

other peaks of granite, still higher than the first: in addition to this, let us imagine in front a gigantic wall, 1500 mètres in height, from the summit of which the Rio Soberbo falls in cascade, afterwards rolling and foaming over the stones in front of the hotel; this wall merging itself on the right in the spur of the mountain already mentioned, and on the left in another mass covered with verdure; and we shall have an idea of the imposing aspect of these mountains.

'A storm at the foot of precipitous mountains of such magnitude is a majestic sight. The summits of the wall and of the granite peaks had disappeared in a thick black cloud. It lightened frequently, and the claps of thunder reverberating in the kind of vat formed by the two spurs of the mountain caused tremendous and continued peals. At the same time torrents of rain, such as are only seen in the tropics, had transformed the road into a river, and the Rio Soberbo into an impetuous torrent; the sound of its cascade mingled with that of the bodies of water which descended from the mountain on all sides, and with the incessant roll and echo of the thunder. The whole formed a scene which must have been witnessed in order to realise its grandeur, but it will be easily understood that it put a stop to our journey.

'On the morrow we started early to climb the mountain. The road we took has been formed on the side of the spur of the Serra, and often skirts immense precipices. The slope of these mountains is too abrupt to allow of their cultivation, but has not prevented a vast virgin forest from establishing itself and completely covering them, as well as the hollow of the narrow valleys between the spurs, which themselves have a considerable inclination towards the main body. In these gigantic crevasses the top of the forest has the appearance of a field dotted with flowers, and in the course of the ascent we passed through a hundred landscapes of indescribable beauty.

'After travelling for two hours we arrived at the village named, after the Empress of Brazil, Theresopolis. It is situated in a pretty, level spot, about 900 yards above the level of the sea, and is overlooked by the rest of the chain and by the summits of the granite peaks. In going from this part towards a hollow on the side of one of these peaks the eye falls upon the gulf of Rio, which, seen from above, appears with its multitude of isles and numerous bays, like a map spread out before the spectator.

'One of our excursions on the mountain took us to one of those beautiful waterfalls which are frequently found in such a country. It was situated in the midst of a thick wood of ancient trees bound with lianas, and fronting a pretty platform where a house might be placed. It presented a delightful prospect. The whole region, in fact, presented at every step harmonious effects of grouping, or beautiful details which arrested the attention.'

To the north of the city of Rio lies the Corcovado, one of the loftiest peaks in Brazil, to which it is customary for the citizens to make excursions for the sake of the extensive prospect it commands. Von Spix thus

CASCADE.

describes it :—' Scarcely were we beyond the streets and the noise of the town when we stopped, as if enchanted, in the midst of a strange and luxuriant vegetation. Our eyes were attracted, sometimes by gaily-coloured birds or splendid butterflies, sometimes by the singular forms of the insects and the nests of wasps hanging from the trees, sometimes by the beautiful plants scattered in the narrow valley and on the gently-sloping hills. Surrounded by lofty cassias, broad-leaved, white-stemmed cecropias, thick-crowned myrtles, large-flowered bignonias, climbing tufts of the mellifluous paullineas, far-spreading tendrils of the passion-flower, and of the richly-flowering hatched coronilla, above which rise the

STORM AT THE FOOT OF THE ORGAN MOUNTAINS.

waving summits of Macauba palms, we fancied ourselves transported to the gardens of the Hesperides. Passing over several streams which were turned to good account, and hills covered with young coppice-wood, we at length reached the terrace of the eminence along which the spring-water for the city is conducted. A delightful prospect over the bay, the verdant islands dispersed in it, the harbour with its crowded masts and various flags, and the city stretched out at the foot of the most pleasant hills, the houses and steeples glittering in the sun, was spread before our eyes. . . . . The stream which the aqueduct conveys to the city falls in one place in beautiful cascades over the granite rocks. Oblique-leaved bignonias, slender costus, and heliconias, the red-flower stems of which shine with peculiar splendour, contrasted with the gloom of the forest; arborescent ferns and grasses, hanging bushes of vernonias, myrtles, and melastomas bending under a load of blossoms, adorn the cool spots that surround them. Large and small-winged butterflies play above the rippling water, and birds of the gayest plumage strive morning and evening to overcome the noise of the stream by their diverse notes. . . . At the cascade, which is called Caryoca, the road turns aside from the aqueduct and leads over a dry eminence, covered with low trees and shrubs, to the forest which clothes the ridge of the Corcovado. The narrow steep path passes over several streams. The vegetation is uncommonly strong and luxuriant; but as we ascend higher the large trees gradually become more rare, and the bamboos and ferns more numerous, among which is a beautiful arborescent fern, fifteen feet in height. When you have made your way through the last thicket you reach the green summit of the mountain, where single shrubs, among which is a magnificent arborescent vellozia, offer to the eye a vegetation resembling that of the higher campos of Minas. From this spot there is a beautiful view, extending over the woods, hills, valleys, and the city, to the sea, the broad surface of which is lost in the distant horizon. Toward the south the mountain is broken, and the prospect loses itself in a steep declivity, bounded by the blue bay of Boto-Fogo; and still further, the bold masses of the Sugar-loaf Mountain close the horizon. At this elevation, of about two thousand feet, the difference in the temperature is already so sensible that you fancy yourself transported to a colder zone. Several streams, flowing from the ridge of the mountain, are always some degrees colder than the water in the aqueduct; and at the approach of sunset, the summit of the mountain is enveloped in clouds, which gradually sink into the valley.'

F

CHAPTER III.

RIVERS, GULFS, AND LAKES.

THERE can be no animal life where there is no vegetation, and no vegetation without water; the sandy plains of Africa and Arabia present a marked contrast to the dense forests of South America. This is owing to the configuration of the land; the vapours raised from the Atlantic by the rays of the tropical sun being driven by the east winds, which prevail during the greater part of the year, against the lofty range of the Cordilleras, where they descend and form the chief supply of those mighty rivers which in length and magnitude surpass all others in the world. These rivers also form the chief means of communication through the forests amidst which they flow, the formation of the land in some parts being such that tribes only two or three leagues apart from each other would require days of travel before they could hold any intercourse.

All the principal rivers flow towards the east. The space between the western side of the Andes and the sea seldom exceeds one hundred miles, and is often less than fifty; consequently, the small quantity of rain which falls on that side has not space to be collected into large streams, but descends in small mountain torrents, violent in their short course, but frequently dried up for a considerable part of the year. Moreover, there are districts of great extent where no rain has been known to fall for centuries. But between the eastern side of the mountains and the coast there are river-basins enormously larger than any in the old world, in which the torrents of rain which fall in the wet season are collected and form streams which unite into a few large rivers, pouring into the ocean a volume of water far exceeding any others. The Irawady, the largest river in India, is 2200 miles long, and in some parts is four miles wide; but its course is slow, it is much encumbered with islands, and the mud it has carried down has formed a delta of considerable extent, through which it reaches the sea by fourteen small channels. The Ganges and the Bramahpootra rise at considerable distances from each other, but unite about forty miles from the coast, and the mud they have brought down, which is estimated at 20,000 tons per second, has formed a delta stretching along the coast for a distance of two hundred miles. And the Indus and the Sutlej also unite, but have no contributory streams, and the delta they have formed extends for a hundred and twenty miles.

But the Amazon is 3700 miles from its source to its mouth; receives the waters of twenty rivers collected from a surface three times the extent of all Europe; it is navigable for more than two thousand miles and the tide rises up it for four hundred; the bed is in some parts six hundred feet deep, and the mouth is a hundred and eighty miles in width; it discharges more water than the eight chief rivers of Asia, and the stream is so powerful that fresh water may be taken up out of sight of land. The La Plata is second only to the Amazon; and the Orinoco, though not so long or so wide as these, is three miles in width at a distance of five hundred and sixty miles from its mouth.

Of these great rivers the Amazon is generally ranked as the largest, though Humboldt considered this as doubtful. Its rise has been attributed to various localities, but the furthest point from its mouth is in the Tunguragua, sixty miles from the Pacific. This stream unites with the Yucayle, which is formed by the junction of two others called the Apurimac and the Paro or Beni. The former of these collects the water of a valley extending two hundred and fifty miles from the great lake of Titicaca to the mountain Knot of Pasco. These sources are one thousand miles apart, and their respective rainy seasons occur at different periods of the year. The valley in which the Apurimac takes its rise is noted for its beauty and fertility, the hills being clothed with luxuriant forests. In the higher parts the stream is rapid, and is interrupted by many falls; but it soon attains considerable magnitude, the breadth at the junction of the Napo being between five and six thousand feet, and the depth six hundred.

As the river flows on it is joined by many tributary streams, the principal of which are the Madeira, the Xingu, the Negro, and the Branca. The Madeira rises among the mountains of southern Bolivia, and in its course of about two thousand miles receives many accessions. Its rise and fall are not simultaneous with those of the Amazon, but take place about two months earlier; so that in December, when the Amazon is lowest, the Madeira is at its height. It is navigable for about four hundred and eighty miles; then occur a series of cataracts and rapids, above which there is another reach of navigable water. The upper branches supply a means of communication with the La Plata and the table-lands of Matto Grosso. Its junction with the Amazon, nearly nine hundred miles from the sea, is described by Mr. Bates as very striking:—'While travelling,' he says, 'week after week along the somewhat monotonous stream, often hemmed in between islands, and becoming thoroughly familiar with it, my sense of the magnitude of this vast water-system had become gradually deadened, but this noble sight renewed the first feelings of wonder. One is inclined in such places as this to think the Paraenses do not exaggerate much when they call the Amazon the Mediterranean of South America. Beyond the mouth of the Madeira the Amazon sweeps down in a majestic reach, to all appearance not a whit less in breadth before, than after, this enormous addition to its waters.' The country about the lower parts of the river is unhealthy, and this circumstance, together with the ferocity of the Indian tribes, has prevented many settlements from being formed on its banks. It was formerly visited by traders who collected turtle, oil, and gums, but hostilities having taken place between them and the Indians it is now less frequented.

The Negro enters the Amazon about eight hundred miles from the sea, and is probably 1000 miles in length. It rises in the eastern Cordilleras, and varies in width from four to twenty miles. A small stream, called the Cassiquiari, forms a communication with the Orinoco, and thus unites that river with the Amazon, and travellers sometimes pass by this means from one river to the other; but the voyage is attended with some danger from the rapids.

These streams flow through gloomy forests and extensive savannahs, the banks being generally lined with trees. For 2200 miles the Amazon is navigable, no cataracts interrupting its course for that distance. As the wind blows constantly from the east it may be ascended at all times, the wind enabling the boatmen to stem the current, which assists them when coming down the river. The tide rises as far as Obidos, four hundred miles from the mouth; and two days before and after the full moon the phenomenon called 'the Bore' takes place. This consists of 2, 3, or 4 waves, 12 or 15 feet high, rushing up with great violence, and, unless proper precautions are used, carrying away or swamping the small vessels employed on the river. There are some parts of the world where it is even more violent: at Tsientang, in China, it rises to the height of thirty feet, rushing up the river at the rate of twenty-five miles an hour, sweeping before it everything that is exposed to its force.

During the wet season the current is rapid, and the country on both sides of the river is inundated for a considerable distance, the water rising in some parts forty or fifty feet; large trees are carried away, and alligators swim where before the jaguar made his haunts. The flooded lands extend for hundreds of miles along the river, varying in width with the formation of the ground. They are called by the natives 'Gapo;' and the Indians show great dexterity in piloting their canoes through the trees, or over the deluged savannahs, being guided by signs which are imperceptible to strangers. For days together the traveller may pass through the tops of trees, now level with the surface of the water, though forty feet above the ground, without ever entering the main river. Sometimes he has to bow his head to avoid the prickly leaves, sometimes they meet in an arch above. Sometimes a level tract covered with grass appears to be beyond the flood, and the traveller unexpectedly finds himself afloat over what had appeared to be solid ground. 'Crossing,' says Mrs. Agassiz, 'to the opposite side of the river (Tapajoz) we came upon a vast field of coarse, high grass, looking like an extensive meadow. To our surprise the boatmen turned the canoe into this green field, and we found ourselves apparently navigating the land, for the narrow boat-path was entirely concealed by the long reedy grasses and tall mallow plants, with large pink blossoms, rising on either side and completely hiding the water below. This marshy, overflowed ground, above which the water had a depth of from four to six feet, was full of life. As the rowers pushed our canoe through the mass of grass and flowers, Mr. Agassiz gathered from the blades and stalks all sorts of creatures: small bright-coloured toads of several kinds, grasshoppers, beetles, dragon-flies, aquatic snails, bunches of eggs—in short, an endless variety of living things. The harvest was so plentiful that we had only to put out our hands and gather it.'

The larger forest-trees are not found on the grounds subject to inundations. The cacao, the sarsaparilla, numberless palms, gigantic grasses, with a variety of shrubs springing up after the wet season to perish on the next overflow, constitute the vegetation of the river-shore. Lianas and other parasites form festoons from tree to tree, and birds of brilliant plumage vie with gay-coloured flowers in giving beauty to the scene.

The second in point of magnitude of the rivers of South America is the La Plata, which is formed by the junction of the Parana and the Uruguay, about two hundred miles from the coast.

The Parana rises in the province of Minas Geraes, has a length of nearly 2000 miles, and is navigable for 1250, being in some parts thirty miles in width. The Paraguay, which is one of its tributaries, rises in the table-lands of Campos Parexis, and flows through a country of most luxuriant vegetation for nearly 1000 miles, receiving in its course the

LIANAS.

Rio Grande and the Pilcomayo, having their sources in the east side of the Andes.

These two rivers bring down an immense body of water, inundating the low country in the wet season to an extent of 48,000 square miles, to the depth of twelve feet, leaving a deposit which greatly promotes vegetation. They contain an immense multitude of fish, and their banks are frequented by many birds, both aquatic and others.

The navigation of the Parana and its tributaries is much interrupted by rapids and falls. One of the branches, the Rio Tiete, which rises on the western side of the coast chain of mountains a few leagues to the south of Rio Janeiro, presents in its course of 400 miles the most extraordinary windings. From Porto Feliz, which is only forty-five leagues from its mouth, the boatmen estimate the distance they have to pass at one hundred and thirty leagues. The vessels employed are made of a single trunk of the iberóva or ximboûva, from fifty to sixty feet long, five feet and a half broad, and three or four deep, carrying a cargo of 400 arrobas, besides provisions. They are usually manned by eight persons, and as their narrow width will not allow the use of sails, are propelled by short oars and long poles.

The river is full of violent currents, rocks, and waterfalls, thirteen of which cannot be passed without landing half of the cargo. The fall at Avanhandavussú, seven leagues from the junction with the Parana, and that at Stapore, are each thirty feet in height, and consequently it is necessary entirely to unload the boat, and forward it and the cargo by land. When the travellers have reached the Parana, the great waterfall of which is three miles further to the north, that river conveys them without hazard, when the dangerous current of Jupia is passed, to the mouth of the Rio Pardo, where they generally arrive on the fifth day. The Parana rolls its immense mass of waters slowly and majestically along in a broad bed, and the navigation on it is agreeable, but attended with some danger when the wind is high, when the shallow boats are beaten by tremendously high waves. The eastern bank is generally high and the western low, consisting of white sand and covered with woods. The Rio Pardo flows, with great impetuosity and considerable fall, through an open country covered with grass. Its navigation is extremely difficult, being interrupted by two-and-thirty cataracts, so that it often requires two months to pass its eighty leagues of course. In the harbour of Sangue-xuga the boats are unloaded, and conveyed on four-wheeled cars, drawn by oxen, to the harbour of Campapuão, at two and a half miles' distance. The crews are often laid up at this place with malignant fevers, caused by incessant hardship and the damp, foggy climate they have passed through. From thence the boats proceed with only half their cargo, till they reach the Rio Cochim. On this river, which winds between steep cliffs and rocks, they have to pass two-and-twenty rapids and falls, at some of which it is necessary to lighten the boat, at others entirely to unload it. From the Cochim they come to the Tacoary, a considerable river, which is generally about seventy fathoms broad, and has only two falls, which are the last of the hundred and thirteen between Porto Feliz and Cujaba. This river comes down with numerous windings through pleasant, grassy plains, into the lowlands bordering the Parana, and empties itself into the main stream by many channels.

All travellers agree in the praise of these countries, where the stranger is constantly surprised by an abundance of new and remarkable objects. The islands and the banks of the river are inhabited by innumerable flocks of birds; incredible shoals of fish come into this river from the Paraguay. Palms of singular forms are seen on the banks, alternating with aromatic grasses and shrubs. When the travellers arrive in the canals between the Pantanaes, thousands of ducks and water-hens rise in the air on the approach of the boats; immense storks wade the boundless swamps; enormous crocodiles inhabit the waters: sometimes the stream flows for leagues together between thick fields of rice, which here grows spontaneously. The diversity and grandeur of the scenery announce the approach to a great river, and the navigators approach the Paraguay, which is of considerable width even in the dry season, but during the rains overflows its banks and spreads into a vast lake above a hundred square leagues in extent. The navigation hence is easy, and the voyage to the mouth of the Rio de San Lorenzo is generally made in eight days; ten days' sail up the Rio Cujaba brings them to the Villa de Cujaba, the whole journey occupying from four to five months.

Opinions vary as to which river should be ranked as third in South America, but the Orinoco is generally considered to hold that place. It rises in the southern part of the mountains of Parime, towards the south of Venezuela, and running west, then northwards, and lastly in an easterly direction, after a course of six hundred leagues, falls into the ocean a little to the south of the Island of Trinidad, forming so strong a current that ships, with all their sails set and a westerly wind in their favour, can scarcely make way against it. The upper part of the river runs through nearly level plains, where the banks are covered with forests, containing a great variety of trees, animals, and birds, but few human beings. The rainfall in the wet season, which commences in the end of April, amounting to two or three hundred inches, these plains are subject to great inundations, the river rising from thirty to thirty-six feet, overflowing the banks, covering the ground for four hundred square leagues, and leaving a deposit of mud which afterwards produces a most luxuriant pasture. During the inundations the inhabitants pass round from place to place by crossing the savannahs, instead of keeping to the bed of the river.

It is joined by the Guaviare, the Meta, and the Apure; the Meta being navigable to the foot of the Andes. The tide rises 300 miles from its mouth, and the water discharged by it is found unmixed at a distance of twenty leagues from the coast.

The Orinoco differs from the Amazon in the position of its cataracts, those in the latter river being situated at a short distance from its source, whereas in the Orinoco they occur after it has run more than a third of its course, and therefore offer a much greater obstruction to the navigation. These cataracts form one of the most remarkable features of the river. We shall, therefore, give an account of Humboldt's visit to them, taken from his own narrative.

He embarked with his companions at San Fernando de Apure, on the 30th of March, and they soon afterwards entered a district inhabited only by jaguars, crocodiles, and chiguires (cavies), and flocks of birds in the air so closely crowded together as to appear like a cloud. The river was bordered by a kind of hedge of *sauso* (a species of cæsalpinia), about four feet high, behind which was a copse of cedar, braziletto, and lignum vitæ. The jaguars, tapirs, and peccaries, had made openings, through which they came to drink, without showing any alarm at the boat. '*Esse como en el Paradiso*,' said the old Indian pilot ; but on watching the animals it was evident that they feared and avoided each other.

The hedge of sauso is in some places at a distance from the water's edge. On the intermediate shore crocodiles were seen basking in groups of six, eight, or ten, supposed to be one male and the rest females. These animals were so numerous in the river that there were seldom less than five or six in view. The Indians said that scarcely a year passed without two or three persons being drowned by them. A young girl of Uritucu saved herself by singular intrepidity and presence of mind, when seized by the jaws of a crocodile. When she felt herself taken hold of she sought the animal's eyes, and plunged her fingers into them with such violence that the pain forced him to let her go, after having bitten off the lower part of her left arm. Notwithstanding the quantity of blood she lost, she succeeded in reaching the shore, swimming with the hand that remained. ' I knew,' she said coolly, ' that the cayman lets go his hold if you push your fingers into his eyes.' Mungo Park's guide in Africa twice succeeded in escaping when seized by a crocodile by the use of the same means. So nature teaches the same lesson in different parts of the world.

On their third day's voyage they passed an island called Isla de Aves, inhabited by thousands of flamingos, rose-coloured spoonbills, herons, and moorhens, displaying plumage of the greatest variety of colour, and so close together that they appeared to be unable to move. They stopped for the night at a place where it was difficult to find dry wood for a fire to keep off the wild beasts, but they were undisturbed till 11 o'clock, when 'a noise so terrific arose in the neighbouring forest that it was almost impossible to close our eyes. Amid the cries of so many wild beasts howling at once, the Indians discriminated such only as were at intervals heard separately. These were the little soft cries of the sapajous, the moans of the alouate apes, the howlings of the jaguar and couguar, the peccary, and the sloth, and the cries of the curassao, the parraka, and other gallinaceous birds. When the jaguars approached the skirt of the forest, our dog, which till then had never ceased barking, began to howl and seek for shelter beneath our hammocks. Sometimes, after a long silence, the cry of the tiger came from the tops of the trees ; and then it was followed by the sharp and long whistling of the monkeys, which appeared to flee from the danger that threatened them. We heard the same noises repeated, during the course of whole months, whenever the forest approached the bed of the river.'

When the natives are interrogated on the causes of the tremendous noises made by the beasts of the forest at certain hours of the night, the answer is, ' They are keeping the feast of the full moon.'

On the 3rd they passed a place where great numbers of manatis are caught. These are found in abundance in the Orinoco below the cataracts, in the Meta and the Apure, and are often ten or twelve feet in length, weighing from five to eight hundred pounds. As the clergy consider this animal a fish, they are much sought after for consumption in Lent.

On the 5th they entered the Orinoco, ' where,' says Humboldt, ' we found ourselves in a country presenting a totally different aspect. An immense plain of water stretched before us like a lake, as far as we could see. White-topped waves rose to the height of several feet, from the conflict of the breeze and the current. The air resounded no longer with the piercing cries of herons, flamingos, and spoonbills, crossing in long files from one shore to the other. Our eyes sought in vain those water-fowls, the habits of which vary in each tribe. All nature appeared less animated ; scarcely could we discover in the hollows of the waves a few large crocodiles, cutting obliquely, by the help of their long tails, the surface of the agitated waters. The horizon was bounded by a zone of forests, which nowhere reached so far as the bed of the river. A vast beach, constantly parched by the heat of the sun, desert and bare as the shores of the sea, resembled at a distance, from the effects of the mirage, pools of stagnant water.'

The commotion caused by the junction of the rivers affected some of the party in the same manner as the motion of the sea. The breadth of the river at this point was 1906 toises at low water, which is increased to 5517 toises, or about eight miles, in the rainy season.

Proceeding to an island in the middle of the river which is noted for its harvest of turtles' eggs, they were astonished, after having for several days past seen nothing but desert shores, to find multitudes collected here, belonging to various tribes, each encamped separately, and the shore being regularly divided into lots apportioned amongst them.

The Indian pilot who had brought them from San Fernando being unacquainted with the cataracts they had to make fresh arrangements for continuing their voyage, and they did not leave Pararuma till the 10th. The navigation was impeded by islands and rapids, and they stopped for the night at the priest's house, it being the first time for a fortnight that they had slept under a roof. The environs of the Mission of Carichana appeared to them to be delightful. 'The little village is situated in one of those plains covered with grass that separate all the links of the granitic mountains, from Encaramada to beyond the Cataracts of Maypures. The line of the forests is seen only in the distance. The horizon is everywhere bounded by mountains, partly wooded and of a dark tint, partly bare, with rocky summits gilded by the beams of the setting sun. What gives a peculiar character to the scenery of this country are banks of rock nearly destitute of vegetation, and often more than eight hundred feet in circumference, yet scarcely rising a few inches above the surrounding savannahs. The same phenomenon seems to be found also in the desert of Shamo, which separates Mongolia from China. Those banks of solitary rock in the desert are called *tsy*. I think they would be real table-lands if the surrounding plains were stripped of the sand and mould that cover them, and which the waters have accumulated in the lowest places. On these stony flats of Carichana we observed with interest the rising vegetation in the different degrees of its development. We there found lichens cleaving the rock; and, collected in crusts more or less thick, little portions of sand nourishing succulent plants; and, lastly, layers of black mould deposited in the hollows formed from the decay of roots and leaves, and shaded by tufts of evergreen shrubs.'

On the following day they reached a part of the river where the bed is full of granite rocks, amongst which they passed through channels not five feet broad, and the boat was sometimes jammed between two blocks of granite. When the current was too strong to be resisted, the rowers leapt into the water and fastened a rope to the point of a rock to warp the boat along.

On the 12th they reached the mouth of the Meta, one of the largest tributaries of the Orinoco, in some parts eighty-four feet deep. The junction of the two rivers is surrounded by rocks and piles of granite, appearing from a distance like ruined castles. Wide sandy shores separate the edge of the water from the forest, amidst which were seen solitary palm-trees standing out against the sky on the top of the high grounds. Canoes are sometimes detained here two days before they can pass the whirlpools caused by the rocks.

On the 13th they landed near the rapids of Tabaje, where they found a few Indians residing. These Indians are very unsettled, and rather than labour in cultivating the ground will feed on stale fish, scolopendra, or worms. They do not paint their bodies, but the faces of the girls were all marked with round black spots, like the patches formerly worn by European ladies.

The river is crossed by a chain of mountains, on which it forms a long series of cataracts. 'Nothing,' says Humboldt, 'can be grander than the aspect of this spot. Neither the fall of Tequendama, nor the magnificent scenes of the Cordilleras, could weaken the impression produced on my mind by the first view of the rapids of Atures and Maypures.' These falls are situated within twelve leagues of one another, and divide the Upper and Lower Orinoco. Above them the country was undescribed before Humboldt's time, and they found in the space of 100 leagues only six or eight white persons. Many fables were current among the Europeans respecting headless men, &c., inhabiting the district, which were chiefly derived from the Indians.

The fall of Atures is surrounded by lofty mountains clothed with thick forests, and savannahs covered with slender plants and grasses like European meadows. On the borders are glens where the rays of the sun scarcely penetrate, and the soil is covered with the plants which frequent damp situations. The rocks near the river are the haunts of flamingos, herons, and other fishing-birds, which look like men posted as sentinels.

The bed of the river in an extent of five miles is traversed by innumerable dykes of rock, forming so many natural dams, the space between being filled with islands of various dimensions, which divide the stream into a number of torrents boiling up over the rocks. A great part of the river appeared dry. Blocks of granite are heaped up, as in the *moraines* of Switzerland. The river runs into caverns, in one of which they heard the water roll at the same time over their heads and below their feet. They were struck with the smallness of the quantity of water to be seen in the channel of the river, the little streams into which it was divided finding a passage between the rocks, and breaking over them in foam.

When the barriers are high, light boats are carried on shore and drawn by the help of rollers, made of branches of trees, to parts where the river is again navigable. When they are only two or three feet high the Indians descend in the boats, and in going upwards they swim on before, fix a rope to one of the points of rock, and draw the boat up. It is often filled with water during this operation; at other times it is broken against the rocks, and the Indians, bruised and bleeding, escape with difficulty by swimming.

They saw at the village of Atures a little boy who had had a singular adventure with a jaguar. 'Two Indian children, a boy and a girl, about eight and nine years of age, were seated on the grass near the village in the middle of a savannah. A jaguar issued from the forest and approached the children, bounding around them; sometimes he hid himself in the high grass, sometimes he sprang forward, his back bent, his head hung down in the manner of our cats. The little boy, ignorant of his danger, seemed to be sensible of it only when the jaguar with one of his paws gave him some blows on the head. These blows, at first slight, became ruder and ruder; the claws of the jaguar wounded the child and the blood flowed freely. The little girl then took a branch of a tree, struck the animal, and it fled from her. The Indians ran up at the cries of the children, and saw the jaguar, which then bounded off without making the least show of resistance. The little boy was brought to us, and appeared lively and intelligent. The claw of the jaguar had torn away the skin from the lower part of the forehead, and there was a second scar at the top of the head. This was a singular fit of playfulness in an animal, which, though not difficult to be tamed in our menageries, nevertheless shows itself always wild and ferocious in its natural state.'

While at Atures they suffered much from mosquitos and other

insects, which so abound on the banks of the river that the first question when friends meet in the morning is, 'How did you find the zancudos during the night? How are we off to-day for the mosquitos?' By which they were reminded of the old Chinese salutation, *Vou-to-hou,* 'Have you been incommoded in the night by serpents?' Humboldt doubts 'whether there is any country on earth where mankind are exposed to more cruel torments in the rainy season.' The lower strata of air, from the surface of the ground to the height of fifteen or twenty feet, are absolutely filled with venomous insects. At one place the inhabitants are accustomed to stretch themselves on the ground and pass the night buried in sand three or four inches deep, leaving out only the head, which they cover with a handkerchief. An Indian said to one of the priests, ' How comfortable must the people be in the moon! She looks so beautiful and so clear, that she must be free from mosquitos.'

Having again embarked they proceeded to Maypures, where they remained three days. The river here runs at the foot of the eastern chain of mountains, having receded from its ancient bed towards the west, where a savannah about thirty feet above the water extends to the cataracts. These are formed by an archipelago of islands and rocky dykes, filling the bed of the river for 3000 toises. From the summit of a mountain at a short distance a sheet of foam is seen extending for a mile. Enormous masses of black stone arise in the midst. Some are rounded peaks grouped in pairs, others resemble towers, castles, and ruins. Their dark hue contrasts with the brightness of the foam. Every rock and islet is covered with clusters of trees. A thick vapour rises over the river, surmounted by the tops of lofty palms, the feathery leaves of which rise almost straight towards the sky. At every hour the aspect varies. Sometimes the hills and trees throw their shadows over it; sometimes the rays of the sun are refracted in the overhanging cloud, and coloured arcs are formed which alternately appear and fade away.

The inhabitants of Maypures are a mild, temperate people, who cultivate plantains and cassava, and manufacture pottery, which they decorate with great skill. In their huts an appearance of order and neatness was found rarely met with in the houses of the missionaries.

Having reached San Fernando de Atabapo, near the junction of the Atabapo and the Guaviare with the Orinoco, they proceeded up the former river, in which they found everything changed—the condition of the atmosphere, the colour of the waters, and the form of the trees. They had suffered much inconvenience from the disagreeable taste and smell of the water of the Orinoco, but they found that of the Atabapo pure, agreeable to taste, and without any perceptible smell. After several days' voyage up the river they saw a rock near the confluence of the Rio Temi, called the *Piedra de la Madre* (the Mother's Rock), and on inquiring the cause of the denomination were told the following tale, which displays the cruelty sometimes exercised towards the natives by the priests employed in the missions :—

'In 1797 the missionary of San Fernando had led his Indians to the banks of the Rio Guaviare on one of those hostile incursions which are prohibited alike by religion and the Spanish laws. They found in an Indian hut a Guahiba mother with three children,— two of whom were still infants,— occupied in preparing the flour of cassava. Resistance was impossible : the father was gone to fish, and the mother tried in vain to flee with her children. Scarcely had she reached the savannah when she was seized by the Indians of the mission, who go to hunt men, like the whites and the negroes in Africa. The mother and her children were bound and dragged to the bank of the river. The monk, seated in his boat, waited the issue of an expedition of which he partook not the danger. Had the mother made too violent a resistance the Indians would have killed her, for everything is permitted when they go to the conquest of souls ; and it is children in particular they seek to capture, in order to treat them in the mission as *poitos*, or slaves of the Christians. The prisoners were carried to San Fernando, in the hope that the mother would be unable to find her way back to her home by land. Far from those children who had accompanied their father on the day in which she had been carried off, this unhappy woman showed signs of the deepest despair. She attempted to take back to her family the children who had been snatched away by the missionary, and fled with them repeatedly from the village of San Fernando: but the Indians never failed to seize her anew ; and the missionary, after having caused her to be mercilessly beaten, took the cruel resolution of separating the mother from the two children who had been carried off with her. She was conveyed alone towards the missions of the Negro, going up the Atabapo. Slightly bound, she was seated at the bow of the boat, ignorant of the fate that awaited her ; but she judged, by the direction of the sun, that she was removed farther and farther from her hut and her native country. She succeeded in breaking her bonds, threw herself into the water and swam to the left bank of the Atabapo. The current carried her to a shelf of rock, which bears her name to this day. She landed and took shelter in the woods ; but the president of the missions ordered the Indians to row to the shore and follow the traces of the Guahibi. In the evening she was brought back. Stretched upon the rock (la Piedra de la Madre), a cruel punishment was inflicted on her with those straps of manati leather which serve for whips in that country, and with which the alcades are always furnished. This unhappy woman, her hands tied behind her back with strong stalks of mavacure, was then dragged, to the mission of Javita. She was there thrown into one of the caravanserais that are called *las Casas del Rey*. It was the rainy season, and the night was profoundly dark. Forests, till then believed to be impenetrable, separated the mission of Javita from that of San Fernando, which was twenty-five leagues distant in a straight line. No other route is known than that of the rivers; no man ever attempted to go by land from one village to another, were they only a few leagues apart. But such difficulties do not stop a mother who is separated from her children. Her children are at San Fernando de Atabapo ; she must find them again ; she must execute her project of delivering them from the hands of Christians—of bringing them back to their father on the banks of the Guaviare. The Guahiba was carelessly guarded in the caravanserai. Her arms being wounded, the Indians of Javita had loosened her bonds, unknown to the missionary and the alcades. She succeeded, by the help of her teeth, in breaking them entirely ; disappeared during the night, and at the fourth rising sun was seen at the mission of San Fernando, hovering around the hut where her children were confined. "What that woman performed," added the missionary who gave us this sad narrative, "the most robust Indian would not have ventured to undertake!" She traversed the woods at a season when the sky is constantly covered with clouds, and the sun during whole days appears but for a few minutes. Did the course of the waters direct her way ? The inundations of the rivers forced her to go far from the banks of the main stream, through the midst of woods, where the movement of the waters is almost imperceptible. How often must she have been stopped by the thorny lianas that form a network around the trunks they entwine! How often must she have swam across the rivulets that run into the Atabapo! This unfortunate woman was asked how she had sustained herself during the four days. She said that, exhausted with fatigue, she could find no other nourishment than those great black ants called *vachacos*, which climb the trees in long bands to suspend on them their resinous nests. We pressed the

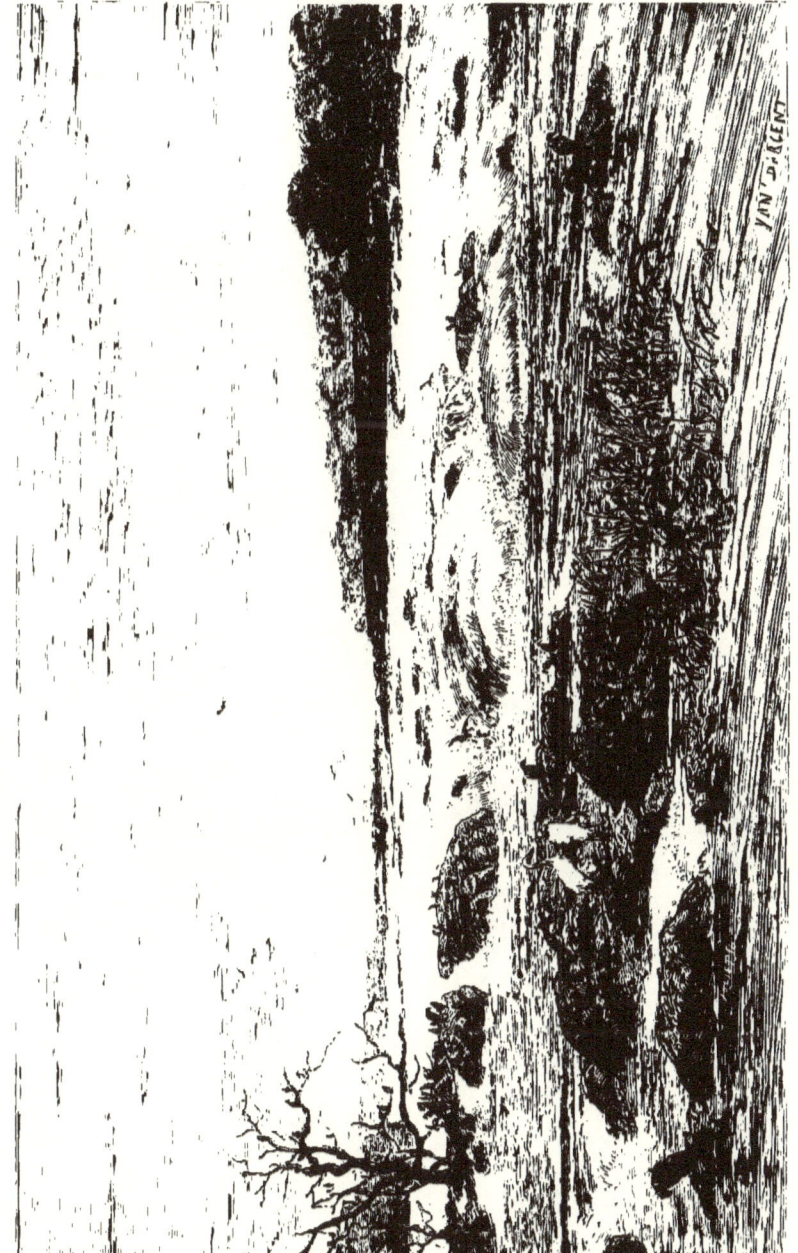

THE RAPIDS OF PIRAPORA.

missionary to tell us whether the Guahiba had peacefully enjoyed the happiness of remaining with her children, and if any repentance had followed this excess of cruelty. He would not satisfy our curiosity ; but at our return from the Rio Negro we learnt that the Indian mother was not allowed time to cure her wounds, but was again separated from her children and sent to one of the missions of the upper Orinoco. There she died, refusing all kind of nourishment, as the savages do in great calamities.'

Having reached the mission of Javita they were detained several days, whilst their boat was conveyed overland from the Tuanimi to the Pimichin, which runs into the Rio Negro, by which they proceeded to the Amazon.

Among the rivers of second degree of magnitude the Magdalena and the San Francisco are two of the chief. The former, which rises in the Knot of Los Papas, in the Central Cordillera, has a course of about 1000 miles in length, one half of which is navigable ; but the voyage is rendered tedious by the violence of the current, and occupies twenty days, sometimes increased in the rainy season to fifty or sixty days. The direction of the valley in which it flows being from north to south, the trade-winds are not felt, and the air is stagnant like that in some parts on the Orinoco. Consequently the heat is excessive, the climate is unhealthy, and the insects superabundant. The plain of Bogota, however, which is situated above the valley of the upper Magdalena at a height of 8000 feet, is temperate, and has two rainy seasons, whereas in the valley below there is but one.

It is joined about two hundred miles from the coast by the Cauca, the upper part of which flows through a valley two hundred miles long and twenty-five miles wide, and then through a narrow gorge for more than a hundred miles, in which it forms a succession of rapids and falls, the valley gradually widening in the lower part till it unites with the Magdalena. The lower part of the Cauca is subject to great inundations, and the lower valleys of both streams are generally covered with grass interspersed with bushes, but few trees.

The San Francisco rises in the southern part of the coast chain of Brazil, and its length is about 1500 miles, flowing into the Atlantic nearly midway between Pernambuco and Bahia. It has many rapids and falls, the principal one at Alfonso being fifty feet in height. Not far from this is one named Pirapora, an Indian name signifying 'Leaping Fish'—rocks rising in the midst of the river, on which fish, leaping out of the water to avoid being carried away, frequently fall and die. Flocks of aquatic birds hover over it or perch on the trees and stones to watch for their prey.

A little to the north of the port of Espiritu Santo is the small river Doce, the banks of which are remarkably beautiful. They are covered with thick forests, which are the haunt of a great number of different animals. 'Here are frequently found the anta, or tapir, two kinds of wild swine, two species of deer, and above seven varieties of the cat kind, among which the spotted ounce and the black tiger (jaguar) are the largest and most dangerous.' The river is navigated in narrow canoes propelled with long poles; as there are many banks which in the dry season appear above water they can always be reached even when it is at its height, but as at that season it is very rapid, at least four men are required to stem the current. The water is then turbid and yellowish, and is said to generate fevers. There are many islands, and the number is said to increase in the higher parts of the river, being covered with ancient trees, and each having its particular name.

The cries of many monkeys are heard from the surrounding forests, particularly the *barbados*, the *sauassus*, &c. Magnificent macaws are seen soaring above the lofty *sapucaya* trees, screaming with loud voices, their glowing red plumage shining with dazzling splendour in the light of the unclouded sun. Paroquets, maracanas, maitaccas, teribas, curicos, camutangas, nandayas, and other varieties of parrots, fly in numerous flocks from bank to bank, and the stately Muscovy duck is stationed on the cecropias on the margin of the river; toucans and çurucuas utter their loud cries. The beautiful bright green crowns of the palmetto are seen amongst the trees, and on the banks below splendid flowers of white and deep yellow form thick, close wreaths about the bushes.

There are not many rivers in Mexico; the Rio Bravo del Norte, sometimes called the Rio Grande—a name which it scarcely deserves—is the largest. It rises on the Sierra Verde of New Mexico, to the east of the Great Salt Desert, and is about 1800 miles long, but is notwithstanding only a mountain torrent. The current is rapid, and the banks steep; but the bed is so shallow that for two hundred miles from its mouth it is navigable only for vessels drawing three or four feet, and those drawing five or six feet cannot, in general, ascend higher than one hundred miles. The lower part of its channel is only about two hundred yards wide, and the mouth is impeded by a bar.

The gulfs of South America are some of the finest in the world.

WAVES BREAKING ON THE SHORE.

That of Mexico, on the north, is 1100 miles in length, and seven or eight hundred wide, occupying an area of 800,000 square miles.

The Gulf of Cariaco, which is supposed to have been formed by some volcanic disturbance in comparatively recent times, is about thirty-five miles long and forty-eight broad. The port of Cumana, which is situated on it, Humboldt considered sufficient to receive all the navies of Europe, is generally from forty-five to fifty fathoms deep, affords excellent anchorage, and is as calm as a lake.

The bay of Todos Santos, which is the port of Bahia, is one of the finest harbours in America, twenty-eight miles in length by twenty in width: it has two entrances; the most eastern is five miles wide. The neighbouring country is exceedingly fertile and productive.

The harbour of Rio de Janeiro is about sixty or seventy miles in circumference, opening to the Atlantic by a deep entrance about a mile in width, with steep rocks on either side; that on the north, Pao d'açucar, being known by shipmasters as a sea-mark. 'Towards noon,' says a German traveller, 'we came up to these colossal rocky portals, and passed between them into a great amphitheatre, in which the water appeared like a tranquil inland lake; and scattered flowery islands, bounded in the background by a woody chain of mountains, rose like a paradise full of luxuriance and magnificence. The banks, in bright sunshine, rose out of the dark blue sea, and numerous white houses, chapels, churches, and forts, contrasted with their rich verdure. Rocks of grand forms rise boldly behind them, the declivities of which are clothed in all the luxuriant diversity of a tropical forest. An ambrosial perfume is diffused from these noble forests, and the foreign navigator sails delighted past the many islands covered with beautiful groves of palms.'

There are few good harbours on the west coast, and the swell of the Pacific makes the entrance frequently hazardous, and sometimes impracticable except by means of the *balsas* made by the natives. 'These are composed of seal-skins inflated; two are generally sewed together end to end, and the balsa is formed by lashing two of these side by side, laying some canes on the top. The man who manages the balsa sits astride on the aftermost part, and impels the balsa with a double paddle, broad at each end, which he holds by the middle; and so dexterous are the natives that there is not the least danger of being upset, or even of being wetted by the surf. On these original, and apparently precarious rafts, all the merchandise is landed at Arica, and all the specie brought to the vessels, except the sea be very calm, and the surf runs low.'

The lakes in South America are not numerous, and are not to be compared to those of North America. That of Titicaca, which has already been mentioned, is the largest; it is about three hundred miles in length and fifty or sixty in width, and of great depth; several small streams flow into it, and its surplus waters are discharged by a small stream into another lake called Pansa, which is about the same length, but so narrow as to appear a mere cleft in the hills, and without any apparent outlet.

The lake of Tacarigua, or Valencia, in Columbia, is larger than the lake of Neufchatel, being about ten leagues in length and two in breadth. It is remarkable from the fact that its waters are gradually diminishing, and in Humboldt's time vast tracts of land were planted with sugar, cotton, or plantains, which formerly were covered with water. Several small streams flow into it, but it has no visible outlet. The environs are unhealthy in the hot season.

There are other small lakes in Peru and elsewhere, but none of any great extent. In Mexico, however, they abound: the chief are Lake Chapalu, in which the two streams forming the Rio Grande del Lorma unite, covering an area of 1300 square miles; and Lake Terminos, which receives the River Lucantin.

CHAPTER IV.

CLIMATE, STORMS, WHIRLWINDS, MIRAGE.

IN a country of such great extent of surface, and such varying elevation, the climate is inevitably variable in a corresponding degree. Accordingly we find that in some parts of South America rain has been unknown for centuries, whilst in other parts it falls in such torrents as to inundate the land over thousands of square miles; in some parts the climate is described as a perpetual spring, while in others summer is equally constant. In some winter is the wet season, in others the rains fall in summer. In some the rainy and dry seasons follow in regular succession, each persisting through its accustomed period; in others the climate is variable to excess, presenting snow, rain, sunshine, dense clouds, and thunderstorms, in rapid succession in the course of a few hours. In going through such a country travellers pass in a few hours from one climate to another. Lieut. Brand, in his account of the passage of the Andes, describes the situation on one day, shut up in a casucha in the midst of a snow-storm, with the thermometer 15° below freezing, and 48 hours afterwards his arrival amongst evergreen shrubs, with peach-trees in blossom, wild flowers in abundance, and the air filled with fragrance. Humboldt says that at Cumana it scarcely ever rains, whilst at Cumanacoa, only seven leagues distant, there are seven months of wintry weather; and that, whilst La Guayra is one of the hottest places on earth, Caracas, at a distance of only about six miles, enjoys the coolness of spring, the thermometer falling at night to 53°.

It is by the dryness or rain, rather than by the temperature, that the seasons are distinguished in the countries near the equator. The variations of the thermometer in any given place are generally much more limited in their range than in temperate regions, and the change of seasons is not from hot to cold, but from dry to wet, or the reverse, and these changes are generally very distinct and regular in each place. Different winds usually prevail during the wet or dry seasons, and they vary very much in different parts. For instance, in the interior of Brazil, the dry season answers to winter, whereas at Pernambuco, eight degrees south of the equator, and on the coast, winter is the rainy season: winter is also the rainy season in Chili; but on the east of the Cordillera it occurs in summer. Towards the south and near the coast, as at Rio de Janeiro, the dry and wet seasons are less distinct than in the interior or at Pernambuco, but yet it is the winter that is generally the driest season in the eastern part of the continent. These variations arise from the prevailing winds, the form of the land, and the more or less wooded character of its surface.

The maximum intensity of the heat also greatly depends on the nature of the soil. In sandy deserts, like those of Africa, the heat reaches its highest point during the day. On the ocean it is not so high, but it is more uniform. In South America, where dense forests abound, the evaporation is great, and the vapour carries off the excess of heat.

In like manner with the temperature, the effects of the light of the sun upon the atmosphere varies greatly with the distance from the equator. And in this also the nature of the soil plays an important part. In countries where large tracts are almost destitute of vegetation, as in a great part of Africa, the air is very dry, and sand raised by the winds deprives it of part of its transparency, from the want of rain to clear the atmosphere. In other parts of the torrid zone, on the Atlantic, on the American continent, in the islands of the South Seas, and some parts of India, vapour in a transparent form is largely mingled with the air; and instead of the bluish-gray colour which it has in our climate and in the sandy deserts, the sky presents a deep blue tint, which gives it a special character in the zenith. Even at the horizon the blue colour, though lighter, is still remarkably distinct, whereas in the temperate zones the shade is always whitish.

In hot climates, the presence of humidity in the air not only gives the sky its deep blue colour during the day, it also adds to the wonders of equatorial nature luminous appearances of incomparable beauty. At sunset, especially, spectacles are presented of a magnificence surpassing description; and the superior beauty of sunset over sunrise is owing to the greater humidity of the atmosphere after the heat of the day, than in the morning, when the coolness of the night has caused part of the vapour to be deposited in dew. It is not on the land that the finest sunsets are seen; but at the same time the celestial blue of the distant mountains, the rose or violet tints which the nearer hills exhibit, shaded by their relative distances, the warm tints of the ground, the deepening shadows of the valleys, while the higher peaks still reflect his rays, when the sun has just disappeared below the horizon, harmonise in a marvellous manner with the brilliant yellow of the sky, with the red or rose tints, the deep blue of the zenith, and the still deeper colour, which, from contrast, often seems greenish, reigning in the east. The

SUNSET AMONG THE MOUNTAINS.

blending of these colours, with the variations of the ground and the rich vegetation of the nearer parts, form graceful pictures, offering valuable suggestions for the artist. Sometimes clouds, light and rose-coloured, or, more dense, fringed with gold or copper colour, produce particular effects, resembling some of our northern sunsets; but when the sky is clear the shadows show a special character, totally different from those of the temperate zone. Sometimes bright fringed clouds intercept the view of the neighbouring mountain-peaks; sometimes the tops of

I

the more distant hills intercept a portion of the sun's light in the upper regions of the atmosphere, and produce the curious phenomenon of twilight rays. Then may be seen, stretching from the point where the sun has disappeared, a number of rays, or rather of large divergent bands of rose-coloured light, reaching sometimes to 90 degrees, and in some cases even to the opposite point of the horizon.

But on the sea, when near the Equator, the visible sky is free from clouds, and when the diverging rays blend with the twilight, the play of lights defies all our powers of description. How is it possible to describe the red and rose-coloured tints of the twilight rays edging the dazzling gold colour of the western sky? How, above all, can we describe the inimitable blue, varying from that of noon, which occupies the part of the sky between the zenith and the twilight? Added to this is the reflection on the water agitated by the trade-wind, the deep neutral tint of the sea in the east, the

CLOUDS OVER THE MOUNTAINS.

white foam of the breaking waves, the pale rose of the eastern sky, and the greenish gloom of the horizon.

Those who have seen these splendid views can never forget them, but words are powerless to describe them.

In the western part of Peru the summer commences in November. The light yellow sands reflect the sun's rays with burning heat. Every living thing which does not quickly escape from their power is overtaken by certain destruction. No plant finds nourishment in the

TWILIGHT RAYS.

arid soil; no animal finds support on it. No birds or insects move in the scorching atmosphere. Only in the highest regions the majestic condor, the king of the air, may be seen, soaring towards the coast. Only where the ocean and the desert meet is there life and movement. Flocks of carrion-crows swarm over the remains of marine animals which lie scattered along the shore; otters and seals move about the rocks, inaccessible to man; hosts of sea-fowl eagerly pounce on the fish and mollusca cast up by the waves; variegated lizards play about the hillocks of sand; and crabs and sea-spiders furrow their way over the moist beach.

In May the scene changes. A thin veil of mist overspreads the sea

CLIFFS.

and the shore, increasing in thickness during the following months till October. At first it rises between nine and ten o'clock in the morning, and disappears about three p.m.; but in August and September it becomes thicker, and continues to hang constantly over the land. It does not resolve into rain properly so called, but descends in a fine, thick fall, which the natives call *Garua*.

Though the garua sometimes falls in large drops, it differs from rain in that it does not descend from clouds at a great elevation, but is formed in the lower regions of the atmosphere by the union of minute bubbles of mist. The average height to which it extends does not exceed 1200 feet, and it never rises above 2000 feet above the sea, and it is found only within a few miles of the coast: beyond its range heavy rains fall; and the boundary line between the rain and the mist may be exactly defined. There are two plantations, one half of which is watered

HURRICANE IN THE ANTILLES.

by the garuas, and the other half by rains, the boundary line being marked by a wall. When the mists set in the appearance of the hillocks bordering the sandy plains is entirely changed. As if by magic, vegetation covers the ground which a few days previously was a barren wilderness. Horses and cattle are driven to graze on these grounds, and find abundance of rich pasture during several months. There is no water, but they do not appear to suffer from the want of it, for they are always in good condition when they leave.

In summer most of the rivers are dried up, and in some of the northern parts, where the garuas are scanty, the vegetation depends entirely on the mountain rains, and when these are deficient the cattle suffer greatly. At Piura there is such a total absence of dew that a sheet of paper left in the open air during the whole night does not in the morning show the least sign of moisture; and in one year, when the rains did not come on at the usual time, fodder was so scarce that one haciendado lost 42,000 sheep. In Central and South Peru, the moisture scarcely penetrates the ground for half an inch.

In the oases the garuas are much heavier than in the adjoining desert. Along the whole of the coast for a considerable extent there is no rain below the level of 7000 feet above the sea.

On the eastern side of the Andes the rainy season generally commences with thunderstorms, exceeding in violence any which are witnessed in the temperate zones. The hurricanes generally commence on the verge of the tropics bordering the temperate zone, in the vicinity of continents and islands, especially in the districts where volcanoes are situated. The neighbourhood of the Antilles, the Gulf of Mexico, the Indian Ocean, the Isle of Bourbon, are the regions where their effects are most destructive, whole plantations being sometimes overwhelmed, houses destroyed, and even the materials of which they were constructed being swept away. On the northern coast of Brazil they are not so common, but at Rio and to the southward they are both frequent and violent. Captain Hall gives a description of one which occurred when he was on the coast:—'The day broke with an unwonted gloom, overshadowing everything; a dense black haze rested like a high wall round the horizon; while the upper sky, so long without a speck, was stained all over with patches of shapeless clouds flying in different directions. As the sun rose he was attended by vapours and clouds which concealed him from our sight. The sea wind, which used to begin gently and then gradually increase to a pleasant breeze, came on suddenly and with great violence, so that the waves curled and broke into a white sheet of foam as far as the eye could reach. The sea looked black and stormy under the portentous influence of an immense mass of dark clouds, rising slowly on the western quarter till they reached nearly to the zenith, where they continued suspended like a mantle during the whole day. The ships which heretofore had lain motionless on the surface of the bay, were now rolling and

BOATS CAUGHT IN A STORM.

pitching, with their cables stretched out to seaward; while the boats that used to skim along from the shore to the vessels at anchor were seen splashing through the waves under a reefed sail, or struggling hard with their oars to avoid the surf breaking and roaring along the coast. The flags that were wont to lie idly asleep by the side of the mast, now stood stiffly out in the storm. Innumerable sea-birds continued during all the day, wheeling round the rock on which the town stood, and screaming as if in terror at the sudden change. The dust of six months' hot weather, raised into high pyramids, was forced by furious gusts into the innermost corners of the houses. Long before sunset it seemed as if the day had closed, owing to the darkness caused by the dust in the air, and to the sky being overcast in every part by unbroken masses of watery clouds. Presently lightning was observed among the hills, followed shortly afterwards by a storm exceeding in violence anything I ever met with. During eight hours deluges of rain never ceased pouring down for a moment; the steep streets of the town soon became the channels of continued streams, of such magnitude as to sweep away large stones, rendering it everywhere dangerous, and in some places impossible to pass. The rain found its way through the roofs, and drenched every part of the houses; the deep rumbling noise of the torrents in the streets never ceased; the deafening loudness of the thunder, which seemed to cling round the rock, became distracting; while the continual flashes of the forked lightning, which played in the most brilliant manner, from the zenith to the horizon on all sides, were at once beautiful and terrific. I never witnessed such a night. As the day broke the rain ceased, and during all the morning there was a dead calm; the air was so sultry that it was painful to breathe it, and though the sky remained overcast, the sun had power enough to raise up clouds of steam, which covered the whole plain as far as the base of the mountains.'

Sometimes during such storms the unusual phenomenon is observed of lightning ascending, either in a line like that seen flashing across the sky in ordinary storms, or spreading out like the leafless branches of a tree. It is well known that during storms there is what is called the return stroke, and instances have occurred of persons being killed by it, but it is seldom that the opportunity occurs of witnessing it. Captain Fitzroy, however, gives an extract from the narrative of the voyage of *La Peyrouse* describing three such; and he also mentions that when an English ship was lying at Zante in 1823, two of the sailors were knocked down by an electric shock, which ascended along the chain-cable by which she was anchored.

The rain which accompanies tropical storms is proportioned to their violence in other respects: it is described by an eye-witness as beginning 'with a few drops large as bullets, falling slowly; followed by the whole mass of water, beating down everything, and forming, in a few moments, a depression in the earth two or three inches deep.'

The hail which falls is of proportionable magnitude. Mr. Darwin, on his journey from Bahia Blanca to Buenos Ayres, was told of stones which had fallen during the previous night 'as large as small apples,' descending with such violence as to kill many of the wild animals. 'One man,' he says, 'had already found thirteen deer lying dead, and I saw their *fresh* hides; another of the party, a few minutes after my arrival, brought in seven more. Now I well know, that one man without dogs could hardly have killed seven deer in a week. The men believed they had seen about fifteen dead ostriches, part of one of which we had for dinner; and they said that several were running about evidently blind of one eye. Numbers of smaller birds, as ducks, hawks, and partridges, were killed. I saw one of the latter with a black mark on its back, as if it had been struck with a paving-stone. A fence of thistle-stalks round the hovel was nearly broken down; and my informer, putting out his head to see what was the matter, received a severe cut, and now wore a bandage.' In India, also, enormous hailstones have been seen, from 6 oz. to 1 lb. in weight; and one mass of ice of more than 1 cwt. is reported to have fallen.

The tubes of vitrified sand, called *fulgurites*, are not unfrequently found in the sandy plains which abound in South America. These tubes, which are formed by lightning entering loose sand, vary from two to four inches in diameter, and have been traced to a depth of thirty feet. The thickness of the tube varies from a thirtieth to a twentieth of an inch, and is occasionally as much as a tenth. The internal surface is vitrified, glossy, and smooth; on the outside the grains of sand are rounded, and have a slightly glazed appearance. Those which Mr. Darwin found near Maldonado were nearly in a perpendicular direction; but one of them deviated considerably from a straight line, and divided into two branches, one proceeding downwards and the other turning upwards. Several were found within a short distance, from which it would appear that the lightning had divided into separate branches before entering the ground. These tubes give an astonishing idea of the force of a flash of lightning, which in an instant could form several such tubes in such a refractory material as quartz.

A storm which occurred during Mr. Darwin's visit to South America produced the following singular effects in a house belonging to the Consul-General at Monte Video :—' The paper, for nearly a foot on each side of the line where the bell-wires had run, was blackened. The metal had been fused, and although the room was about fifteen feet high, the globules, dropping on the chairs and furniture, had drilled in them a

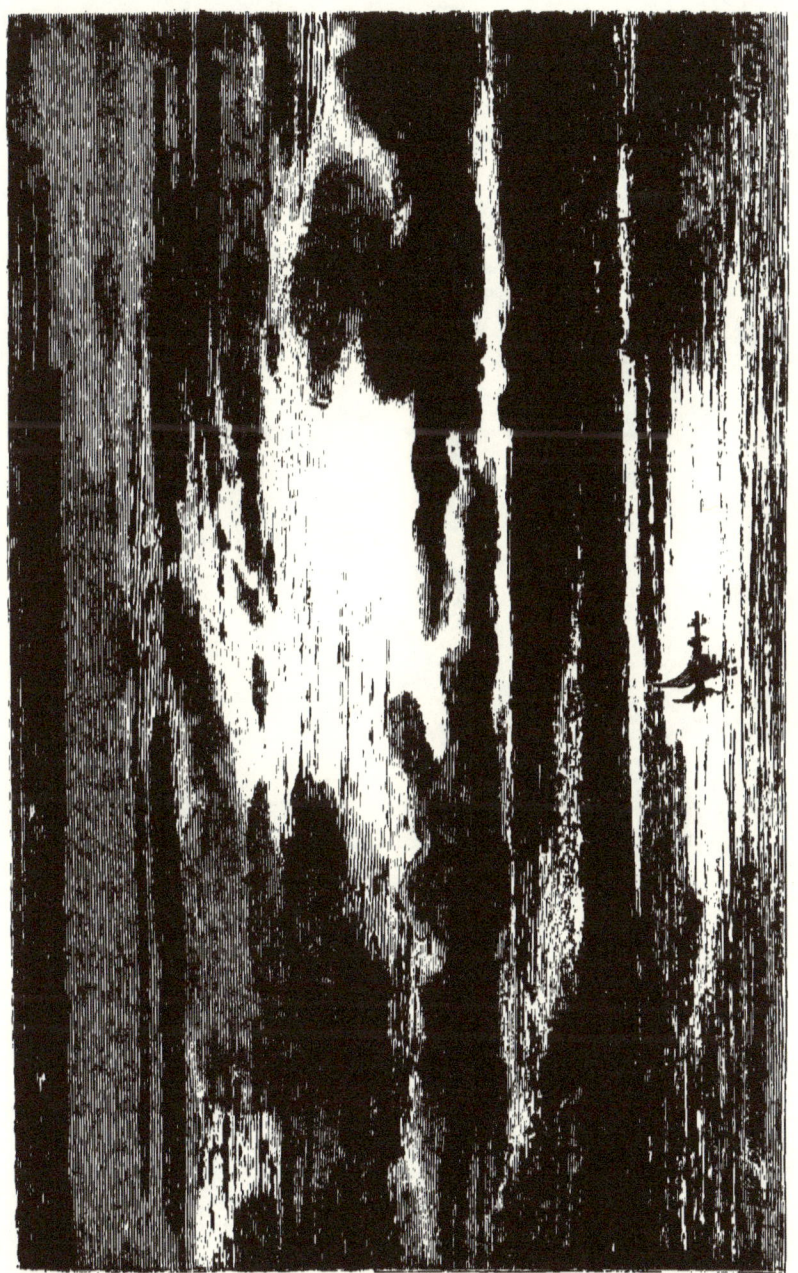

ARBORESCENT LIGHTNING.

chain of minute holes. A part of the wall was shattered as if by gunpowder, and the fragments had been blown off with force sufficient to dent the wall on the opposite side of the room. The frame of a looking-glass was blackened, and the gilding must have been volatilized, for a smelling-bottle, which stood on the chimney-piece, was coated with bright metallic particles, which adhered as firmly as if they had been enamelled.'

A tropical storm in the dense forests of South America is one of the grandest scenes conceivable. It begins with a roaring in the upper regions of the atmosphere, which descends lower and lower; the crashing of the trees as their branches are driven one against another, or even their trunks uprooted from the ground, mingling with the thunder pealing through the forest, and the cries of the disturbed animals, combine to form an uproar not easily imagined.

Sir Francis Head gives a ludicrous account of a storm which surprised his party as he was crossing the Pampas. Thinking that the weather looked threatening, he got into the carriage to sleep; some of the party laid down below, and the rest on the ground in different directions. 'About midnight we were awakened by a most sudden and violent whirlwind, which blew several of the party's clothes away, and they were afterwards found in the river. There was so much dust that we could scarcely breathe, and all was utter darkness until the lightning suddenly flashed over our heads; the thunder was unusually loud, and down came a deluge of rain. The wind, which was what is termed a "Pampero," was now a dreadful hurricane, and I expected every moment that it would overturn the carriage. I sat up and looked round me, and in my life I never saw so much of the sublime and of the ridiculous mixed together. While the elements were raging, and the thunder was cracking and roaring immediately above us, the lightning would for an instant change the darkness to the light of day. In these flashes I saw our party, who were all hallooing one to another, in the most ludicrous situations. Some were lying on the ground, afraid to sit up, and holding their ponchos and clothes, which were trying to escape from them; some, who had lost their clothes, were running half-naked towards the post-room; others had lost their way, and were standing against a dead wall, not knowing where to go. A French Colonel, who had travelled in the carriage from Mendoza, was lying on a stretcher made of a bullock's hide, grasping his clothes, which were now wet through, and vociferating at his cowardly servant, who, instead

K

of assisting him, was standing about ten yards from him crossing himself. In vain did he call him in Spanish every sort of "animal;" the fellow, who had been approaching his master, was riveted to the ground by the unexpected sound of the church-bell, which, from the violence of the hurricane, occasionally gave a solitary toll. The rain beat so violently into the two-wheeled carriage, and it shook so terribly, that its inmate could bear it no longer, and in his shirt he ran through the rain. At last they all got into the post-room, and as I looked out of the window of the carriage I saw them all crowded together, peeping over each other's heads at the door.'

Contradictory accounts are given of the climate of Rio de Janeiro. Mad. Pfeiffer complains of it as oppressive, Mr. Darwin describes it as delightful. The difference is probably owing to the different times of the year at which they visited it. Mr. Darwin, on the other hand, did not find the climate of Chili agreeable, while other writers describe it as being the best in the world. The climate of the table-lands is generally equable, but generally misty, except in some parts of Mexico and Peru.

When the wet season has commenced the obscurity of the forests is rendered still more gloomy. Thick fogs veil the sun by day and the moon by night; a brown shade hangs over the depths of the forest, and the stars are invisible. Still, this season has beauties of its own. Thousands of organized beings then awake, and revived by the combined effects of heat and moisture, exhibit all the activity of which they are capable. Innumerable frogs make their voices heard from the pools of water and the overflowed marshes ; the loud tones of a bull-frog are also heard, surprising a stranger who does not know whence they proceed; paroquets fly screaming from side to side, to keep their damp wings in motion. Branches lying on the ground rotting with age, and hollowed out by slow decomposition, are inhabited by a multitude of insects. The leaves of the plants, and many brightly-coloured flowers, open out to the refreshing rain, and seem to acquire new life. The draconthium, the pothos, the bromelia, the cactus, the epidendrum, the heliconia, and a multitude of others, lift up their heads, and many of them exhale the sweetest perfumes. And when the rains are passed away the woods resume all their beauty ; where a few days before various shades of green met the eye, flowers of the richest colours are seen, and birds, whose plumage vies with them in brilliancy, flutter among the blossoms and dart from branch to branch.

Travelling in the rainy season is inevitably attended with both discomfort and danger. Von Spix on one of his journeys says, 'We travelled almost constantly enveloped in a thick fog; for several days together the thermometer, morning and evening, was 57°, and it rose only a few degrees at noon. The numerous mountain-streams had overflowed their banks to a great distance, the roads were broken up by them, the bridges carried away, and the low grounds suddenly converted into lakes . . . . . The mules could scarcely proceed on the bottomless roads; we were forced either to wade or swim through the overflowed streams which we had to pass. If, in the evening, we met with an open shed, or a dilapidated hut, we had to spend the greater part of the night in drying our wet clothes, in taking our collections out of the chests, and again exposing them to the air. Often we had not even the comfort of resting ourselves round the fire, because the wet wood emitted more smoke than flame. In this gloomy wilderness we met with but few huts, chiefly inhabited by mulattos; and besides milk and black beans no kind of provision was to be found.' Another traveller, when he arrived at the Pasamayo, found that it had overflowed its banks. 'Several travellers had stretched themselves on the ground to wait for the morning. No Chimbadores (guides who conduct travellers across rivers) were to be had: my negro guide looked at the water with dismay, and declared that he had never before witnessed so furious a swell. However, we had no time to lose, and I resolved to attempt the passage. Trusting to my well-tried horse, which had already carried me safely through many difficult coasting journeys, I cautiously rode into the river, which became deeper at every step. The overwhelming force of the stream was felt by my horse; and he presently lost his footing, though he still continued to struggle vigorously against the force of the current. At this juncture some passing clouds obscured the moon, and I lost sight of a group of trees which before leaving the opposite beach I fixed my eye upon as a guide-mark. Quite powerless, my horse and I were carried away by the stream, and driven against a rock in the middle of the river. I now heard the anxious outcries of my negro and the travellers on the bank, whilst the waves rose over my head. With a convulsive effort I pulled the bridle, and the horse then turning completely round, once more gained solid footing. I then gave him the spur, and the courageous animal, dashing again into the midst of the current, swam with me to the bank. I rode forward with my negro in search of a better fording-place, and after several fruitless attempts we at length found one, and crossed the river in safety. The other travellers did not venture to follow our example, but called begging us not to leave them behind. I sent the negro back on my horse to bring them over, and the noble animal went backward and forward no less than seven times without making one false step.'

The phenomena called whirlwinds, which seem to have a peculiar relation to storms, are not uncommon in some parts of South America. They have often been confounded with other atmospheric disturbances, such as the tornados which occur near the equator, and the cyclones or great circulating storms of various parts of the globe. But these are very different, and it is important that they should be clearly distinguished.

The whirlwind consists of a column of air two or three yards in diameter, moving in a rotatory manner, and rising gradually to the height of a hundred yards or more. It often appears suddenly in the midst of a plain in calm weather, and without any previous indication of its formation; but sometimes the heated air may be seen quivering over the parched ground, forming in the distance an imperfect mirage. Mr. Belt says that in Australia he has often seen two in action at the same time in different parts of the same plain. They generally move slowly in an horizontal direction, raising the dust and dry leaves, which render their motion visible, like great columns of smoke. They seldom last many minutes, nor do they cause much damage.

The columns of sand raised in the Sahara are caused by whirlwinds. They also occur in the temperate regions; but it is on the parched plains of tropical countries that they are most frequently seen. M. Liais gives a description of two which he witnessed in the course of a few days. The first was about noon on the 1st of October. He was crossing a plain near the Rio San Francisco, when suddenly he heard the sound of rustling of dry leaves, and on looking in the direction whence it proceeded, he saw a column of dust moving in a rotatory manner, about fifty yards to the left of the path he was taking, and approaching the road a short distance in front of the party. 'I urged my horse forward,' he says, 'that I might meet the whirlwind, which I succeeded in crossing. I had in my hand a small white parasol, such as is often used when travelling under the vertical sun of the climate. As soon as I reached the edge of the column I felt the parasol forcibly drawn towards its centre. In trying to hold it, I was almost thrown from my horse, and only withdrew it in a torn state. After the whirlwind had passed over me I saw it become higher, and soon disappear at a short distance.'

'The second occasion was two days afterwards at Porto das Andorinhas. We were on the left bank of the river. On the opposite side was a wide space of open ground surrounded by large trees. I suddenly saw in the midst of this open space the dry leaves scattered on the ground move a short distance, and then remain still for a few seconds. Soon they were disturbed afresh, and suddenly they began to rise with the dust, whirling round, and forming a column which continued to increase in height. It took the direction towards the river, which it soon reached. I could then see clearly that the water slightly rose,

WHIRLWIND.

foaming, in the middle of the column, and a small cloud of water, like dust, was drawn into the whirlwind, which gradually ceased in three or four minutes. The surface of the river was disturbed for only a short distance around. The column passed about sixty yards from us, and there was a slight breeze scarcely to be felt as it did so. Moreover, the air was perfectly calm at the time when it commenced and after it had ceased, and the sky was clear as in the former case, only it was slightly whitened by the smoke which is always caused by the large fires made in the campos at this time to burn the grass. It was also accompanied, like the former, with a peculiar rustling sound; probably caused, in part at least, by the dry leaves drawn up.'

In like manner with whirlwinds, waterspouts often rise in calm weather, but they are always surmounted by large storm-clouds. They consist in a cone of mist, generally reversed, with the broad part towards the cloud, but sometimes it is towards the earth. The diameter, even at the smaller end, is generally larger than that of whirlwinds, and varies from twenty to a hundred yards. When it occurs on land it overturns the houses and buildings over which it passes; it uproots trees, and

WHIRLWIND.

generally carries away everything which is found within its range. They have been seen to carry away all the water of lakes over which they have passed. The trees which they throw down often appear as if struck by lightning. All their sap is evaporated, and they are split into strips. These phenomena seem clearly to indicate the action of strong electric currents. Moreover, they are accompanied by a peculiar sound like that of the passage of electricity through an imperfect conductor, and frequently a storm bursts from the cloud on which they rest. When they occur on the sea, the water is often raised; it is sometimes depressed at the centre of the waterspout, but then it is raised at the circumference.

This terrible phenomenon is probably owing to the same cause as whirlwinds; and if that cause is electric the large storm-clouds often present may explain the great force they exhibit, the electricity which

WATERSPOUT.

would be expended in a storm of some duration being expended in a few minutes.

The cause of whirlwinds and waterspouts is still uncertain. Some have supposed them to result from the conflict of contrary winds, but their occurrence when little or no motion in the air was previously perceptible shows that opinion to be untenable. Mr. Belt, who has had many opportunities of observing them in different countries, considers them to be occasioned by ascending currents of air arising from the heated surface of the ground, and acquiring a rotatory motion in the same manner as the water assumes a spiral movement when a small opening is made in the bottom of the vessel which contains it. M. Liais attributes them to electricity. Whether any of these opinions

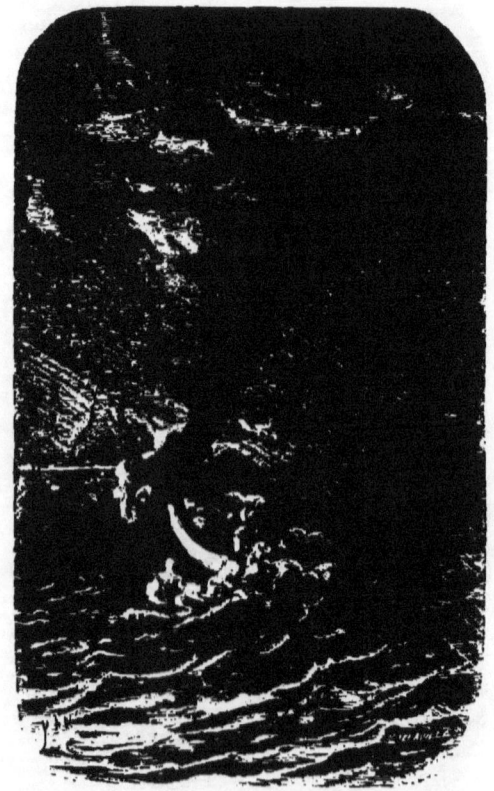

WATERSPOUT.

is correct, or whether the electricity which is often present may not be developed by the pressure of the ascending current, remains to be ascertained by further observation.

Connected with the subject of whirlwinds are the Medanos. These are chains of small hillocks, which intersect the plain of sand extending from the foot of the Andes on the western side to the coast, and which greatly increase the danger of journeying over the plain. The stormy winds raise the sand in columns of from 80 to 100 feet high, which whirl about in all directions, as if moved by magic, and sometimes they suddenly overwhelm the traveller, who only escapes from them by rapid riding.

The medanos have sometimes a firm and sometimes a shifting base. The former are always crescent-shaped, from ten to twenty feet high, with a pointed crest; the inner side being perpendicular, and the outer or bow side rising sharply and forming an acute angle with it. When driven by the wind they pass rapidly over the plains, the smaller and lighter being quickly overtaken by the larger ones, and both being broken by the shock. They assume all sorts of extraordinary shapes, and sometimes move in rows in the most intricate combinations, completely hiding whatever is in the distance. A plain is often covered with a row of medanos, and some days afterwards it resumes its uniform level aspect. Persons of the greatest experience are often misled by them and mistake their way.

The medanos with fixed bases are formed round the blocks of rock which are scattered about. The sand is driven against them by the wind, and when it reaches the top it falls down on the other side, till that is also covered, and a conical hill is thus formed. The small chain of hillocks which intersects the coast obliquely from east to west, checks the progress of the movable medanos, which would otherwise convert fruitful oases into barren flats of sand. A careful observation of these hillock-chains affords the best means of ascertaining the prevailing direction of the wind. On the south they are covered with vast quantities of sand drifted there by the mid-day gales; there is much less on the north side, though not steeper than the south. If one at a distance from the sea runs parallel with the Andes, *i.e.* from S.S.E. to N.N.W., the western side is almost free from sand, as it is driven to the plain below by the south-east wind, which constantly alternates with that from the south.

These movements are most frequent and most extensive in the hot season, for then the parched sand yields to the slightest breeze. In the cold season it absorbs moisture, and is consolidated, more easily resisting the wind.

The state of the atmosphere which precedes whirlwinds is also favourable to the production of the appearance called *mirage*. This phenomenon, which is produced by the varying density of the air, whether caused by heat or moisture, presents different aspects according to cir-

L

cumstances. Sometimes objects appear to be removed from their real position, as in a case recorded to have occurred in Kent, in which a castle appeared to the spectator to be on the side of the hill nearest to him, whereas it was really on the further side. Sometimes they seem to be reflected in a lateral direction, as in another case, when a young lady, who had ascended a mountain in Wales with some friends from whom she became separated, was surprised by the appearance near her

MIRAGE.

of a figure, which she did not for some little time discover to be her own reflection, it being also apparent to her friends, who asked her when she rejoined them who had been her companion. At other times the image appears in an inverted position, either singly or in addition to the original. Captain Scoresby on one occasion distinctly recognised his father's ship at sea by its inverted image in the air, although the distance between the two ships was as much as thirty miles, and the ship was, therefore, far below the horizon of that from which it was observed.

Humboldt describes the appearance of the mirage in the Llanos, on one occasion, when 'the little currents of air that swept the surface of the soil had so variable a temperature that, in a drove of wild oxen, one part appeared with the legs raised above the surface of the ground while the others rested on it. A well-informed person assured us that he had seen, between Calabozo and Urituca, the image of an animal inverted without there being any direct image. Niebuhr made a similar observation in Arabia. We several times thought we saw on the horizon the figures of tumuli and towers which disappeared at intervals, without our being able to discern the real shape of the objects. They were perhaps hillocks, or small eminences, situated beyond the ordinary visual horizon.'

The extraordinary brilliancy of the tropical sunsets has already been

THE SETTING MOON.

spoken of; another astronomical appearance, which is seen with great distinctness in tropical regions, is that called by the ancients *lumen incinerosum*. When, under a clear sky, the moon is observed when only two or three days old, we may distinguish the whole surface of her disc, though only a narrow arc of bright light appears. The other part of her surface appears faintly illuminated, or as if a dark veil covered it. This light, which the ancients called *lumen incinerosum*, translated by the French, *lumière cendrée*, has been supposed by some to proceed from a kind of phosphorescence of the moon, only perceptible when the side which is turned towards the earth is only partially illuminated by the sun. It is, however, now generally held to proceed from the light which the moon receives from the earth being reflected back, and thus rendering her visible. This light is the greatest at the time of new moon, because it is at that time that the earth, as seen from the moon, would appear fully illuminated by the sun's rays; but as at that time the moon rises and sets with the sun, the light diffused through the atmosphere by the latter prevents the *lumière cendrée* from being perceptible: accordingly it is two or three days after the new moon that it appears to be the most intense; but its actual intensity has really declined, because it does not then receive so much light from the earth as at the time of the new moon. It is in the tropics that the light is seen most distinctly, owing to the shortness of the twilight in those regions, and the great transparency of the atmosphere.

The phenomenon called the Zodiacal Light is also witnessed near the tropics, with a brilliancy unknown to those who have only observed it in the temperate zones. This appearance, which in Europe is only seen in the winter months, was unknown to the ancients, and the first notice of it was published about the middle of the seventeenth century. It is seen soon after sunset in the months of January, February, March, and April, and before sunrise in November, in the form of a band of light of a conical shape, stretching from the horizon upwards in an oblique direction, and has generally been supposed to consist of a flattened spheroid, filled with very attenuated matter and surrounding the sun. But on approaching the equator this light loses its conical form, and appears as a band of light entirely encircling the heavens, and remaining visible the whole of the night. Those portions which are nearest to the sun are of about equal brightness with the Milky Way; the more distant parts are fainter, and are visible only in consequence of the extreme clearness of the tropical atmosphere when free from clouds.

This appearance when clearly seen, as in the tropics, is one of the most beautiful of celestial phenomena. It is of the purest white. It has been supposed by some to have a reddish tinge, but that was probably owing to European observers having seen it mingled with the last rays of sunset. In the months of July and August under the tropic of Capricorn, and in January and February for the tropic of Cancer, it is seen after sunset, perpendicular to the horizon. It then rises as a beautiful upright white column, the centre of which is brighter than the brightest part of the Milky Way. The edge of the column is gradually blended with the light of the sky, in which it differs from the Milky Way, some parts of which show a sharp outline contrasted with the ground of the sky.

In the tropics this beautiful sight attracts the attention of the most

THE ZODIACAL LIGHT.

indifferent to celestial phenomena, whereas in Europe many centuries passed before its existence was suspected. Stars of the sixth and seventh magnitude are lost in its light, which allows only those of more than the fifth magnitude to be seen. Its light is more feeble in the zenith, and close attention is requisite to follow its track, which is altogether lost in the Milky Way, and reappears below as a pale and uncertain gleam.

It is the most brilliant when the air has been cleared by heavy rain. In Europe it is not seen in summer, owing to its inclined position and the length of the twilight; but near the equator the shortness of the twilight permits it to be seen all the year round. Still there are periods when it is brightest, which depends upon the position of the sun with regard to the ecliptic.

CHAPTER V.

VEGETATION, FORESTS.

LUXURIANCE, variety, and richness of colouring, are characteristics of tropical vegetation. Mexico is said to contain 'the whole scale of vegetation,' to be 'the rendezvous of the floras of all lands:' the forests of Guiana excited the enthusiasm of Waterton; those bordering on the Orinoco that of Humboldt; and all travellers in Brazil unite in celebrating the variety and beauty found in those which clothe the high lands of that empire.

The atmosphere which surrounds our globe consists, to the extent of one part in 3000, of carbonic acid, and this minute portion is the original source of all the vegetation on its surface. Light, heat, and moisture, are necessary to enable the plant to absorb this element and convert it into a portion of its own substance, and the character of the vegetation of a country is chiefly dependent upon the conditions under which it is subjected to the action of these three agencies.

Vegetation, therefore, varies according to the nature of the climate, and that of the soil most general in any locality. Where, like the Sahara of Africa, or the Dahna of Arabia, it consists of sandy plains at a distance from the sea, exposed to the rays of a vertical sun, no rain falls, it remains, with the exception of a few rare oases, destitute of water, no tree nor shrub can find support, and it is inhabited by neither man, animal, nor insect. The Llanos and Pampas of South America, subject alternately to the burning heat of a tropical sun and to the downpour of tropical rains, are covered at one time of the year with dust, and at the other with rank grass and thistles; the Tell of Africa, subject to the same alternations, produces prolific crops after the annual rains; the Campos of South America present to the eye and the ear many charms for both senses, and the Silvas of the Amazon, more favoured than either, are covered with dense forests, through which the sun's rays scarcely penetrate, and inhabited by an inconceivable variety of animated creatures.

The degrees of heat and moisture in any locality are greatly modified by the prevailing winds, and these by the form of the land. The vapours raised from the Atlantic by the rays of the sun are carried by the trade-winds across the high lands of Brazil, and, condensed by the snow-crowned tops of the Andes, fall in snow or rain, and feed those mighty rivers which serve to fertilise as well as form channels of communication through that empire. But from the western side of the Cordillera to the coast, between lat. 4° S. to lat. 32° S. is nothing but sand, on which no rain falls, and consequently utterly barren, except where short and rapid mountain torrents help to fertilise the soil. Northwards of 4° S. the trade-wind is less regular, torrents of rain fall periodically, and the land towards the west coast is covered with luxuriant growth. And south of lat. 38° S. where the gales bring moisture from the Pacific, every island on the west coast is covered with dense forests, while on the east of the Cordillera the plains of Patagonia show only a scanty vegetation.

The distribution of plants is chiefly regulated by temperature; as we pass from the equator to the pole the vegetation of each zone has its own distinct character. In this distribution each climate has its own peculiar beauty, which is recognised by the painter's expressions, 'Swiss scenery, Italian scenery, &c.' The same kind of change is apparent on ascending lofty mountains, and while in tropical countries the low lands near the coasts are inhabited by the plants peculiar to hot climates, these disappear, and are succeeded in turn by those of the temperate and arctic regions as the domain of perpetual snow is approached.

The influence of temperature in localising vegetation is shown by the fact that 'there are certain plants for which the summer of the tropics, notwithstanding its high temperature, is not long enough to mature the seed, and which prefer the neighbourhood of the equator. Thus the cocoa-nut, which is found in abundance at Bahia, does not succeed at Rio de Janeiro; the trees are only found on some islands in the bay, where the nuts are only imperfectly ripened. The mango also yields excellent fruit at the former place, whilst at Rio, though it attains a considerable height, it bears little fruit, and only in certain years. That magnificent tree, which is originally from India, and is justly reputed one of the finest productions of the hot regions of the earth, abounds in the environs of Bahia. Their immense rounded heads, the thickness and deep colour of their foliage, their deep shade, which no ray of the sun can penetrate, render the groves formed of them in the public gardens of Bahia always cool and agreeable, even in the hottest part of the day. In this garden the jack-tree is intermingled with the mango. This curious bushy tree has its enormous fruit attached to the trunk. The bread-fruit-tree of Tahiti is also abundant at Bahia. The large leaves, incised at the edge in a curious manner, its size, and its fruit, render this tree very ornamental. And the cacao-tree is another of the majestic and valuable productions of this province.' 'The landscape in the neighbourhood of 'Bahia,' Mr. Darwin says, 'almost takes its character from the mango and the jack-trees. Before seeing them I had no idea that any trees could cast so black a shade on the ground.'

The influence of climate on vegetation is shown in the substitution of certain genera for others which occupy the same position in temperate climes. 'The family of Rosaceæ gives to the north its pears, its apples, its peaches, its cherries, its plums, its almonds; in short, all the most delicious fruits of the Old World, as well as its most beautiful flowers. The trees of this family by their foliage play a distinguished part in the vegetation of the temperate zone, and impart to it a character of their own. The Myrtaceæ give to the South its guavas, its petingas, its araças, the juicy plum-like fruit of the swamp myrtles, many of its nuts, and other excellent fruits. This family, including the Melastomaceæ, abounds in flowering shrubs, like the purple queresma, and many others not less beautiful; and some of its representatives, such as the Sapucaia and the Brazilian-nut tree, rise to the height of towering trees. Both of these families sink to insignificance in the one zone, while they assume a dignified port and perform an important part in the other.'

Another illustration of the diversity of vegetation caused by climate is found in the variations in the species of those plants which are common to the temperate regions. In the more northern parts the Euphorbiaceæ are small and unobtrusive, but on the Amazon they assume the form of colossal trees, constituting a considerable part of its strange and luxuriant forest growth. The giant of the Amazonian woods, the Saumaumera, whose majestic flat crown towers over all other trees, while its white trunk stands out in striking relief from the surrounding mass of green, is allied to our mallows. Some of the most characteristic trees of the river shore belong to these two families.

The variety of both animal and vegetable productions on the globe is thus connected with the diversity of climates, arising from the mode of distribution of the sun's rays. But climate seems to be only indirectly the cause of this variety, and not itself to have produced it; that is to say, it is not the change of climate which occasions the diversity of forms, but that beings possessing a certain organization can only con-

THE MANGOE AND THE JACK TREE.

tinue to exist in a climate adapted to that organization. In fact, both animals and vegetables are continually being transported from one climate to another, and whilst some continue to live and to increase, preserving their generic and specific characters, either with or without modification, others refuse any such modification and perish; which proves that the former had no real acclimatization to submit to, but that their organization was originally adapted to the new climate. Even the specific characters are preserved intact. It was formerly believed that the different colours of different races of man arose from difference of climate. It has now been shown that the European retains the characteristics of his race under the equatorial sky, and that the African transported to America, where the natives are yellowish, or to Europe, where they are white, remains as black as in his own country. The sun, no doubt, darkens the skin of men who are constantly exposed to its rays, but it is sufficient to be withdrawn for a time from their influence for this effect to disappear. Nor do the children of white people, resident in a hot climate and embrowned by the heat, show any darker hue when born than those who first see the light in a temperate climate. Moreover, the domestic animals which have been transported to America, and have become wild, have retained their characteristics. The reported return of species to a single type, supposed to be primitive, is a fact which does not appear to have been conclusively verified.

As to plants, there are many mistakes prevalent as to the influence of climate. One may be referred to respecting the castor-oil plant. Treatises on botany describe this plant as woody in hot climates, whilst it is only annual in temperate regions. This is not strictly correct. In hot countries the Ricinus is, in the first year, exactly similar to that which grows in Europe. It yields its seed, but it does not, as in a more northerly zone, die when it has borne fruit, and thereby it clearly differs from annuals. The true annual is the plant which necessarily dies when it has perfected its seed; like the daisy. Whatever care is taken of it after it has fructified, it dies all the same. The Ricinus, on the other hand, continues to vegetate after having yielded its fruit, still throwing out roots. It would flower afresh, and yield fresh seeds, as in the tropics, if the cold of the European winter did not kill it. But this accidental death, as it may be called, does not constitute a change in the nature of the growth of the plant, or a modification owing to climate. It is a casualty which may be prevented by placing the plant in a conservatory to preserve it from cold during the winter. Then it may live during a second, or even a third year, as in hot countries, and presents exactly the same character. It becomes woody at the base and in the lower branches, but it is always, as in the tropics, green in the new branches. At Cherbourg a plant lived in the open air through a winter when there had been no frost, and budded afresh the following year, presenting exactly the appearance just described. Therefore it is not in any respect changed by the climate. The same observation applies to mignonette and other plants exhibiting variations of a similar character.

The countries situated near the equator receiving the rays of the sun nearly perpendicularly, the vital forces in the two kingdoms of organized beings are fully developed. Not that there is no variety of seasons in these countries; they are distinct and clearly defined: but as the temperature is high through the whole year, so the vital energies are in exercise through the whole year. 'When, under the deep blue sky of the tropical regions, the sun's rays strike on the masses of dark green foliage of the forests, and the shadows almost disappear from view, the sheet of light which is spread over the scene affords the spectator a most brilliant prospect. The large butterflies of those climes flutter by thousands round the flowers, whose bright colours they rival. The sensitive plants then open their leaves widely to absorb the carbonic acid of the atmosphere, and their peculiar property of closing at the slightest touch of an insect, or movement of the air, is most highly developed. The reptiles, too, seem to be roused to greater activity, and show a degree of animation to which they are strangers under a colder sky. Caution, however, is needful in attempting to examine their beautiful colours. Clouds of dragon-flies of various hues flit across the water, while swarms of swallows dart in pursuit of the insects which fill the air.

'The aspect of these splendid views when the power of the sun's rays is greatest, the thousand forms of life which give them animation, is very different from the notion of the ancients respecting these regions, which they designated the *torrid* zone, considering them as burnt up by the sun's heat; but wherever there is humidity in the atmosphere, the rays of the sun, far from being destructive, stimulate life.'

But while temperature is the chief agent in stimulating the growth and governing the distribution of plants, it is not the only one; and the range of some plants is limited by circumstances which are only imperfectly known. Thus South America is especially the land of palms, and the cactus is exclusively American; while the ericas are especially African, and though they extend far to the north in Europe, only one indigenous

VIRGIN FOREST IN BRAZIL.

species is known to exist in America. The European and American violas are distinct; some species of rhododendrons, magnolias, azaleas, &c., are found on the east of the Rocky Mountains, but not on the west; some calceolarias are found on the west of the Cordilleras, but not on the east; a new species of palm was found by Bonpland about every fifty miles; each river has its peculiar flora; and while 'there is great resemblance between the tropical vegetation of the East and the West, there is hardly any identity of species—the general aspect is exceedingly different.'

Under the influence of abundant heat, light, and moisture, the vegetation of South America is not only more profuse than that of any other part of the world, but its flora 'is so varied as to render it impossible to convey an idea of its peculiarities, or of the extent and richness of its woodlands.' Its profusion is stimulated by the heat and moisture which prevail over a considerable portion of its surface, and its variety by the different elevation of different parts, presenting variety of climates.

All travellers who have visited the interior of Brazil unite in celebrating the beauty of its forests. 'The untravelled European,' says one writer, 'has not the faintest conception of their magnificence, nor is it possible for any words to give a description of the scene corresponding with the sensations it excites.' Humboldt says:—'When a traveller newly arrived from Europe penetrates, for the first time, into the forests of South America, he beholds nature under an unexpected aspect. He feels that he is not on the confines, but in the very centre, of the torrid zone; on a vast continent where everything is gigantic—mountains, rivers, and the mass of the vegetation. If he feels strongly the beauty of picturesque scenery, he can scarcely define the various emotions which crowd upon his mind; he can scarcely distinguish what most excites his admiration—the deep silence of these solitudes, the individual beauty and contrast of forms, or that vigour and freshness of vegetable life which characterise the climate of the tropics. It might be said that the earth, overloaded with plants, does not allow them space enough to unfold themselves.' To which we may add Mr. Darwin's testimony:— 'Delight is a weak term to express the feelings of a naturalist who, for the first time, has wandered by himself in a Brazilian forest. The elegance of the grasses, the novelty of the parasitical plants, the beauty of the flowers, the glossy green of the foliage, but, above all, the general luxuriance of the vegetation, filled me with admiration. A most paradoxical mixture of sound and silence pervades the shady parts of the wood. The noise from the insects is so loud that it may be heard even in a vessel anchored several hundred yards from the shore, yet within the recesses of the forest a universal silence appears to reign. To a person fond of natural history, such a day as this brings with it a deeper pleasure than he can ever hope to experience again.' And again, on another occasion, he says :—' In England any person fond of natural history enjoys, in his walks, a great advantage, by always having something to attract his attention ; but in these fertile climates, teeming with life, the attractions are so numerous that he is scarcely able to walk at all.'

The wanderer in these forests enjoys refreshing coolness under their deep shade, at the same time that his eye is regaled with the luxuriance of the vegetation. Not only do the trees shoot up to a majestic height, but their trunks are concealed under a thick clothing of verdure ; and if the orchideæ, the pipers and the pothoses borne by a single courbaril, or American fig-tree, were carefully transplanted, they would cover a vast extent of ground. These are so thickly interlaced together that it is impossible for the eye to penetrate between them, and many bear flowers of the most exquisite beauty. And instead of the uniform aspect of European forests, especially in the northern parts, here is the utmost variety in the forms of stems, and leaves, and blossoms. Every moment presents some new object to engage the attention. Even the rocks are covered with lichens and cryptogamous plants of innumerable kinds; and beautiful ferns hang like feathered ribbons from the trees. Some of the colossal trees are so lofty that a fowling-piece would not carry shot far enough to reach the birds perched on their tops, and of such girth that 12, 15, or 18 Indians are unable to encircle them with their outstretched arms. Each of these giants of the forest, which are so closely crowded together, is distinguished from its neighbours in the general effect. ' While the silk-cotton tree (Bombax ceiba), armed with strong thorns, spreads out its thick arms at a considerable distance from the ground, and its leaves are grouped together in light and airy masses, the lecythis and the Brazilian anda shoot out at a lower elevation branches profusely covered with leaves, which unite in verdant arcades. The jacaranda attracts the eye by the lightness of its double-feathered leaves ; the large gold-coloured flowers of this tree and the ipé dazzle by their splendour contrasted with the dark green of the foliage. The spondias arches its pinnated leaves into light oblong forms. A very peculiar and most striking effect in the picture is that produced by the trumpet-tree among the other lofty forms of the forest. The smooth, ash-grey stems rise, slightly bending, to a considerable height, and spread out at the top into verticillate branches standing at right angles, which have at the extremities large tufts of deeply lobated white leaves. The contour of the tree appears to indicate at once hardness and pliability, stiffness and elasticity, and affords the painter a subject equally interesting and difficult for the exercise of his pencil. The flowering cæsalpina, the airy laurel, the lofty geoffræa, the soap-trees with their shining leaves, the slender Barbadoes cedar, the ormosia with its pinnated leaves; the tapia, or garlic pear-tree, so called from the strong smell of its bark; the maina, and many others yet undescribed, are mingled confusedly together, forming groups agreeably contrasted by the diversity of their forms and tints. Here and there the dark crown of a Chilian fir among the lighter green appears like a stranger

TROPICAL VEGETATION.

amidst the natives of the tropics, while the towering stems of the palms, with their waving crowns, are an incomparable ornament of the forests, the beauty and majesty of which no language can describe. If the eye turns from the proud forms of these ancient denizens of the forest to the more humble and lower which clothe the ground with a rich verdure, it is delighted with the splendour and gay variety of the flowers. The purple blossoms of the rhexia, profuse clusters of the melastoma, myrtles, and erigenia; the delicate foliage of many rubiaceæ, and ardisieæ with their pretty flowers, blended with the singularly-formed leaves of the theophrasta; the conchocarpus; the reed-like dwarf-palms; the brilliant spadix of the costus; the ragged hedges of the marantha, from which a squamous fern rises; magnificent stiftia; thorny solana; large flowering gardenias and coutareas entwined with garlands of mikania and bignonia; the far-spreading shoots of the paullinias, of the dalechampeas, and the bauhinia, with its strangely-lobated leaves; strings of the leafless milky bindweed (lianes), which descend from the highest summits of the trees, or closely twine round the strongest trunks and gradually kill them: lastly, those parasitical plants by which old trees are invested with the garment of youth; the grotesque species of the pothos and arums; the superb flowers of the orchideæ, the bromelias, the tillandsias, hanging down like *Lichen pulmonarius;* and a multiplicity of strangely-formed ferns: all these combine to form a scene which fills the European naturalist alternately with delight and astonishment.' The lianas both creep on the ground and climb the tops of the highest trees, passing from one to another, and forming such an intricate network that the botanist is often unable to distinguish one from another, and confounds together the flowers, fruits, and leaves, of different species.

Among these creeping plants a bauhinia is very remarkable; its strong woody branches always grow in alternate arcs of circles; the concavity of each arc is as artificially hollowed as if worked by the tool of a cabinet-maker, and on the opposite convex side is a short blunt thorn. This singular plant, which might easily be mistaken for an artificial production, climbs to the tops of the highest trees. Its leaf is small and bi-lobed. Many of the creeping plants have a very strong smell, some agreeable, some the contrary; one smells like cloves, another like garlic. Many of them throw out branches downwards, which, when agitated by the wind, inflict severe blows upon the unwary traveller; after a time they take root, and form a barrier which must be cut away before any progress can be made. Some of them are two feet in circumference, and are over-run by other creepers.

'If we attempt to analyse the features of a virgin forest of America, we shall find that its general mass is composed of dicotyledonous trees belonging to many different families. Amongst these are trees belonging to the leguminous order. Some of these attain a prodigious size. Martius mentions one that fifteen Indians with outstretched arms could only just embrace. It was eighty-four feet in circumference at the bottom, and sixty feet rather higher up. The leaves are often very large, formed by the union of an immense number of delicate leaflets, forming a shady foliage, generally of a bright green, and of a graceful appearance quite unknown in northern forests. Then there are a multitude of malvaceæ, bombaceæ, euphorbiaceæ, and bignoniaceæ; bushy trees with large palmated leaves, sometimes downy, sometimes glossy, giving a thick shade. Next is a large group of myrtaceæ, lauraceæ, terebinthaceæ, ficoidales, malpighiaceæ, &c., with lanceolate leaves and dark shining verdure, producing a mass of impenetrable foliage.

LIANAS.

A fourth group is of the melastomaceæ, one of the most striking in the forest, with tall straight trunks, large rounded heads; also with lanceolate leaves, but much larger than the preceding, with long, strongly-marked veins, of a bright green, and whitish below, with large handsome flowers at the ends of the branches, of rose-colour, white, and violet, mingled together. There are also other kinds, with large, delicately-marked leaves, or with large fleshy leaves.

'To get a good idea of the appearance of a tropical forest, we must think of these characteristic forms of vegetation, with their endless varieties mingled together in every possible manner, with the trunks and branches of these gigantic trees covered by a multitude of parasites, some with flowers of brilliant colours, some eccentric, like those of the epiphyte orchids; others with delicate leaves, like the thousands of ferns, or large and fleshy, like the arads; with beautiful capitals formed by multitudes of palms, sometimes shooting up and carrying their crowns to the upper sphere of the forest, like the cabbage-tree; sometimes with short trunks and thick foliage, like the acrocomia, mingled with a multitude of plants, each of which has a distinct character—canes musaceæ, agaves, piperaceæ, bignonias, &c., filling the lower part of the forest. As soon as we rise a little above the level of the sea, we must add to these the tree-ferns and a variety of other forms, and then think of the whole as filled up with lianas of twenty different families, and place it on a soil varied like that of Brazil, and we shall get an image of a virgin forest and its incredible variety, of which our European woods, with their miniature trees, cannot furnish the slightest idea.'

Beautiful as the forest is in itself, its beauty is enhanced to the naturalist by the multitude and variety of its inhabitants. Monkeys frolic through the verdant arches, chasing each other from branch to branch and tree to tree. Squirrels scamper up and down, as if unable to rest for joy. Coatis gambol among the fallen leaves, or rival the monkeys in their nimbleness. Pacas and agoutis roam about, ready to take flight at the slightest alarm. The sloth, disturbed by the general movement, seeks a spot where he may find quiet and repose. The tiny deer, little larger than a lamb, snuffs the air and fearlessly bounds along, knowing that no enemy is near.

Birds of brilliant plumage fly amongst the trees. The lonely trogon calls mournfully to her absent mate. The rattling of the woodpecker sounds from some lofty branch. 'Tucáno, tucáno,' comes loudly from some tall tree, where the toucans feast on the thickly-clustered fruit. The mot-mot utters the quick cry from which he gets his name. Tuneful thrushes ring their alternate notes like the voice of a single bird. Parrots are chattering, paroquets screaming. Manakins are piping in every tree without rest. Wood-pigeons, startled, take flight, and pheasants of many kinds go whirring through the wood. Tiny creepers, in gay livery, run up the thick tree-trunks, stopping their busy search now and again to watch the intruding strangers. And, most beautiful of all, the humming-birds, like living gems, dart by; now stopping to kiss the sweet-scented flowers, now furiously battling some rival humble-bee. Butterflies of the richest blue, large as one's hand, flutter past, and from the flowers above comes the noisy hum of myriads of gaily-coloured insects. The lizard, in his gorgeous coat of green and gold, starts from his hole in the bank of sand, stopping every instant, with raised head and quick eye, watching for the appearance of danger; and armies of ants, in their busy toil, swarm by without ceasing.

Except at midday, when all creatures seek shade and repose, every hour calls into action a different race of animals. About two hours before daybreak the red monkey begins moaning as though in deep distress; the houtou—a solitary bird only found in the thickest recesses of the forest—begins his low and plaintive cry about an hour afterwards; the tinamus joins with his whistle about the same time; the tree-frogs and toads, with their high and deep notes, and the monotonous chirp of the grasshoppers and locusts, are heard soon afterwards. Then come the squirrels and monkeys, the hoccos and the pigeons, and many other birds of singular plumage and voices—some harsh and loud, others soft and melodious—and the green, blue, and red parrots assembled on the tops of the trees, or flying towards the islands and plantations, fill the air with their screams. The chattering manakins and the screaming orioles, the woodpecker with his sounding blows, and, above all, the deep tones of the campanero, or bell-bird, sounding from the tops of the trees like the strokes of the hammer on the anvil, and appearing more or less distant according to the position of the bird, fill the wandering stranger with amazement.

As the sun's rays gain power the forest voices die away, and from eleven o'clock till three scarce a sound is heard, except the distant notes of the campanero and the pi-pi-yo; oppressed by the heat, the birds retire to the thickest shade and await the coolness of the evening. The few sounds which are heard from time to time rather intensify the feeling of solitude than impart any feeling of life, and the silence soon becomes oppressive, when perhaps a startling shriek bursts forth which seems to pierce the brain—some feeble cavy or helpless monkey has been seized by a stealthy ocelot. When twilight approaches the sounds of the morning are renewed for a time; the monkeys, chattering or hissing, seek their quarters for the night, and the screaming parrots return in flocks to their usual roosting-places. The vampires, bats, and goatsuckers, come from their retreats to chase the moths and beetles along the banks of the rivers; the crickets chirp from sunset to sunrise, while hoarse bull-frogs 'make night hideous' with their deep croak.

The marvellous effects of moonlight spreading over the thousands of wreaths formed by the lianas can be realised only by one who has travelled on a fine night through one of the forests of tropical America. Sometimes the overhanging arch of verdure is so thick that the traveller is plunged in profound darkness, whilst on one side a ray of light finds its way through a gap, strikes upon the elegant form of a tree-fern, or the large velvety leaves of some bignonia, or is stopped by the thick tufts of marantas. In some places where the cassia, the cæsalpinia, or other large trees of the leguminous order predominate, their foliage, closed during the night, allows the light to filter through to the lower range of shrubs of all sorts, on which are shed the straggling rays, dimmed by their passage through the upper part of the forest. At other times in a clearing the spectator perceives a number of curious plants—palms, musaceæ, caladiums, agaves, dracænas, the large leaves of which, plainly seen in the nearer parts, gradually disappear in the deepening shades of the gigantic trees around. Elsewhere the traveller climbs a high point where the air is perfumed with the odour of vanillas, and whence he looks through a vista on the upper masses of foliage strongly illuminated. The glossy leaves of the ficoids and of the guttiferæ, the shining spikes of the various kinds of palms, rising above the general level, here and there reflect a bright ray of light, in strong relief upon the deep black of the shadows. And when the forest extends into a region really mountainous, where peaks of granite allow asclepias, bromiliaceas, and amaryllæ, and often large trees, to overlook the scene —when foaming cascades descend on all sides, then the harmony of the prospect under the soft glow of the moonlight affords inexpressible delight to the spectator.

The very luxuriance of vegetation in these forests tends to decay,—rich and fertile as is the soil, it is not inexhaustible; hence these gigantic trees are engaged in a constant struggle for existence, and impede each other's growth more than the trees in more northern climates. Even those which have attained a considerable height feel the influence of more powerful neighbours, which abstract the nourishment from the soil,

TREE FERN.

and obstruct the genial influence of the sun's rays, and thus render them an easier prey to destructive influences. Ants eat off the bark, and mole-crickets devour the roots; monkeys, parrots, and other birds attack the fruit, the sloth and other quadrupeds devour the leaves and roots. The lianas, too, which form such a characteristic feature of these regions, aid in the work of destruction; many of them fasten upon the trunk of a tree, wind like a cable about it from root to topmost branch, absorb its juices, and thus destroy the source of their own existence, well earning the name they have received of *mata-pao* (tree-killer). M. Agassiz 'saw a lofty palm-tree completely overpowered and stifled in the embrace of an enormous parasite. So luxuriant is the growth of the latter that you do not perceive, till it is pointed out to you, that its spreading branches and thick foliage completely hide the tree from which it derives its life; only from the extreme summit a few fan-like palm-leaves shoot upwards, as if trying to escape into the air and light. The palm cannot long survive, however, and with its death seals the doom of its murderer also.' It is the astonishing number of lianas and parasites which gives the forests of America their peculiar character, and distinguishes them from those of Africa, India, and Australia.

On the low shores in hot climates of all parts of the world is found the mangrove (Rhizophora). It forms close thickets down to the water's edge, so dense as to intercept the rays of the sun, and, preventing the exhalation of putrid miasmata, renders those neighbourhoods excessively unhealthy. Roots drop from the lower branches, which penetrate into the mud beneath, thus forming new stems, and accumulating all sorts of drift, and binding it together, gradually extend the ground. The seeds begin to germinate before they drop, and from their peculiar form float unevenly, so that when they touch the soil the waves as they rise and fall gradually wash the heavier end, which bears the roots, into the soil, and it proceeds to germinate, growing very rapidly, and soon surrounding itself with a series of arcades like its parent: against these sand, mud, and shells collect, and thus many islands are formed. The plant is evergreen, and the wood is very solid; the small descending roots being perfectly straight and very light, are frequently used by the Indians for arrows. The groves formed by it are much frequented by herons and other birds seeking for crabs and fish which are found among them, but it is very difficult to obtain them, if shot, as it is impossible to penetrate through the thicket, or to venture upon the mud beneath when the water has ebbed. Dampier's description of this plant is both quaint and graphic :—' The red mangrove groweth commonly by the seaside, or by rivers or creeks. . . . It always grows out of many roots about the big-

MANGROVES.

ness of a man's leg, some bigger, some less, which at about 6, 8, or 10 foot above the ground, joyn into one trunk or body, that seems to be supported by so many artificial stakes. Where this sort of tree grows it is impossible to march by reason of these stakes, which grow so mixed one among another, that I have, when forced to go through them, gone half-a-mile, and never set my foot on the ground, slipping from root to root. The timber is hard, and good for many uses; the inside of the bark is red, and it is used for tanning of leather very much all over the West Indies.'

South America is especially the land of palms. The Tamanaquas have a tradition that after the Mexican *age of water* the race of human beings was renewed from the fruit of the palm. Palm-trees form the shade and chief beauty of many of the forests. The fruit of some of the varieties is a chief article of food to many of its people; some of them furnish valuable wood; one species furnishes both food and lodging to a whole tribe; and the various valuable products supplied by this order of plants are so many as to remove much of the apparent extravagance of the assertion of Linnæus,—that the region of palms was the first country of the human race, and that man is essentially *palmivorous*.

PALM-TREES.

Some palms are dwarf, only a few feet in height; some reach 150, 180, or 200 feet; some grow on the coast, others on sandy plains; some on the sides of mountains, to the height of 5000 feet above the sea;

some grow singly, some in numerous clumps or groves; but almost all are most prolific in flower and fruit. 'A single spathe of the date contains about 12,000 male flowers; Alfonsia amygdalina has been computed to have 207,000 in a spathe, or 600,000 upon a single individual; while every bunch of the Seje palm of the Orinoco bears 8000 fruits.' The well-known cocoa-nut is the fruit of a palm, millions of which grow on the coasts of Ceylon. Sago is produced by another; the date, which is the chief article of food of the Arabs, is obtained from a palm; the bud of another is eaten, which is commonly known as the 'cabbage' palm. The betel-nut, so much used in India, is produced by the areca catechu, a palm,—oil, wax, sugar, salt, thread, rope, baskets, and multitudes of other useful articles, are made from its pith, its bark, or sap, or exuded from its leaves.

The timber of the Carnauba palm is strong and durable, and is commonly used for the rafters of houses; the leaves make a durable covering for roofs, and will last for twenty years. The fruit is used for food, after being boiled, and the kernel is given to cattle. Thread and cordage are prepared from its fibres. The leaves are two feet in length, and when full grown expand to the same width. If cut from the tree as soon as full grown, and placed in the shade to dry, small light-coloured scales are formed on their surface, which, when collected and heated, melt like wax, and are formed into candles.

The Mauritia grows abundantly on the delta at the mouth of the Orinoco, and a tribe of natives, called Guaranis, construct their habitations on platforms supported by their trunks, or suspend spreading mats from tree to tree, on which they live during the rainy season. These are partly covered with clay, and the women light fires on them for cooking. The tree also supplies them with a sago-like meal, which is dried in thin slices, and constitutes their chief article of food; while the sap when fermented forms a sweet wine. From its many useful products this tree was called by Gumilla '*arbol de la vida*' (tree of life).

Another palm-product is the substance called 'vegetable ivory.' This is obtained from a tree which grows on the banks of the river Magdalena, called by the natives Tagua, or Cabeza de Negro (Negro's head).

One kind of palm climbs up the trunks of other trees; the stem of another bulges out midway, somewhat in the form of a skittle; while another is supported by roots growing above ground, so that in old trees a person can stand upright beneath the stem of the tree.

PALM TREES ON THE SHORE.

One of the most remarkable vegetable productions of South America is that called the *palo de vaca*, or 'cow-tree,' also sometimes called *arbol del leche*, 'the milk-tree.' Having heard of this tree, Humboldt and his companion visited it and found it to be a fine tree, with oblong pointed leaves, and a somewhat fleshy fruit, containing one or sometimes two nuts. When an incision is made in the trunk, a thick, glutinous, milky fluid issues abundantly from it, perfectly free from acrimony, and having an agreeable smell. It is drunk by the negroes and free people who work in the plantations, and the travellers took a considerable quantity of it without any injurious effect. They were assured by the overseer that the negroes grow sensibly fatter during the season when it is most abundant. When exposed to the air, this juice presents on its surface a yellowish, cheesy substance, in layers, which in five or six days become sour and afterwards putrefy.

The cow-tree appears to be peculiar to the Cordillera of the coast, especially from Barbula to the Lake of Maracaybo.

'Among the many curious phenomena,' says our traveller, 'which I observed in the course of my travels, I confess there are few that have made so powerful an impression on me as the aspect of the cow-tree. Whatever relates to milk or to corn inspires us with an interest, which is not merely that of the physical knowledge of things, but which is connected with another order of ideas and feelings. We can hardly imagine how the human species could exist without farinaceous substances, and without the nutritious fluid which the breast of the mother contains, and which is appropriate to the condition of the feeble infant. The amylaceous matter of the cereal plants—the object of religious veneration among so many ancient and modern nations—is distributed in the seeds and deposited in the roots of vegetables; while the milk which we use as food appears exclusively the product of animal organization. Such are the impressions which we receive in early childhood, and such is the source of the astonishment with which we are seized on first seeing the cow-tree. Magnificent forests, majestic rivers, and lofty mountains clad in perennial snows, are not the objects which we here admire. A few drops of a vegetable fluid impress us with an idea of the power and fecundity of nature. On the parched side of a rock grows a tree with dry and leathery foliage, its large woody roots scarcely penetrating into the ground. For several months in the year its leaves are not moistened by a shower; its branches look as if they were dead and withered; but when the trunk is pierced a bland and nourishing milk flows from it.

It is at sunrise that the vegetable fountain flows most freely. At that time the blacks and natives are seen coming from all parts, provided with large bowls to receive the milk, which grows yellow and thickens at its surface. Some empty their vessels on the spot, while others carry them to their children.'

One of the greatest ornaments of these forests is the Sapucuya, or pot-tree, a species of lecythis, which grows to upwards of 100 feet in height, spreading into a majestic crown, which is extremely beautiful in the spring, when the rose-coloured leaves shoot out and the large white blossoms open. It bears a large nut about the size of a child's head, the top closed with a lid so loosely attached that when the weight of the fruit turns it downwards it separates, and the seeds fall out. These are collected in large quantities by the Indians, who are extremely fond of them, and eat them either raw, or preserve them roasted and pounded; the shells are used as drinking-cups. At the season it is dangerous to remain in the woods, especially when there is any wind, on account of these heavy nuts falling from so great a height.

One of the same family is the Bertholletia, which produces the well-known Brazil nuts. The fruit is as large as a cocoa-nut, with a shell nearly half an inch in thickness, within which the triangular seeds are so closely packed that, when once removed, they cannot be replaced. This shell is exceedingly hard and tough, and tries the skill of the monkeys, who are very fond of the seeds. A friend of Mr. Edwards related to him an amusing incident of which he had been witness: a monkey, pounding one of these nuts on the ground to break the shell, clumsily brought it down with tremendous force on the tip of his own tail, when down dropped the nut, and away flew the monkey, bounding and howling fearfully. When the nuts are fresh they much resemble the cocoa-nut in taste, and the juice, easily expressed, forms a substitute for milk.

The cacao-tree, from which chocolate and the cocoa now so much used are obtained, is another product of these regions. The tree is about fifteen or twenty feet high, with yellowish green leaf, in shape somewhat like that of the birch, and a small red flower, the fruit growing directly from the trunk or branches. The berry or pod is about eight inches in length and five in diameter, containing a white acid pulp, in which are embedded from thirty to forty seeds. From these the chocolate and cacao are prepared. They were extensively used for this purpose in Mexico in the time of Montezuma, and it was so much esteemed by Linnæus that he named the tree Theobroma—'food for a

A FOREST OF ARAUCARIAS.

god,' to mark his sense of its excellence. The seeds were formerly used as money, six of them passing for about the value of a halfpenny.

The Algaroba, or carob-tree, is of such utility that a law was made to prevent its being cut down. It is about the size of an oak, the wood very hard, and bears numerous clusters of pods, about four inches long, containing five or six black seeds like small beans. When ripe the pod is of a brown colour, and has a sweet taste: the cattle are very fond of it, and become very fat; the mules that are fed with the pods return from a journey of a hundred and forty leagues in good condition; but its chief use is as food for large numbers of goats. These animals reach the lower branches of the tree themselves, and they are afterwards assisted by the goatherds, who climb the trees and beat down the leaves and pods with long canes. At certain times of the year, when the pods become scarce, the goats will follow the goatherds without the need of a driver, as if conscious that their provision depended on the help of their keeper. Some of them become so plump that it is not uncommon for one goat to yield a hundred pounds weight of fat, which is all separated from the flesh, the latter being considered of very little value. The goatherds are very superstitious, and believe that some men have power, by witchcraft, to convey the fat of one flock of goats to another if precautions against it are not used; and for the prevention of this they have various charms, which they fasten to the necks or horns of the old goats, especially those which are called the captains of the flocks. These Indians asserted to Mr. Stevenson that a flock of fat goats had been placed under the care of an unskilful herdsman, and that in one night a *hichesero* (wizard) had deprived them of all their fat, and conveyed it to another flock, to the astonishment of the guardian, who in the morning found his fat flock reduced to skin and bone, bleating their lamentations for the loss they had sustained.

The seringa-tree, from the juice of which the waterproof shoes are prepared, also grows here; and the papaw-tree, which 'has the singular property of rendering the toughest animal substance tender, by causing a separation of the muscular fibre: its very vapour even does this. Newly-killed meat suspended among the leaves, and old hogs and old poultry, when fed on the leaves and fruit, become tender in a few hours.'

The Araucaria is scarcely to be considered a tropical production, but belongs to the southern temperate zone, vast forests of this noble tree are however found in the chain of Mantiquiera.

Cassava, which formerly constituted the chief sustenance of the native Indian tribes of South America, is produced from the root of two varieties of a plant called Jatropha, the Jatropha manihot, and the Jatropha janipha, called in Brazil mandioc. The former of these is in its natural state highly poisonous, but its injurious property is destroyed by exposure to heat. Even the poisonous juice, which proves destructive to cattle or poultry, if drunk by them shortly after it has been extracted, forms a favourite soup, called by the Brazilians *casscripo*, and is found wholesome and nutritious when boiled with meat and seasoned.

PAPAW-TREE.

The roots somewhat resemble parsnips, and are generally about fourteen or fifteen inches long, and four or five inches thick in the middle. In preparing them they are first peeled, and then pressed against a revolving wheel, which reduces them to pulp. This is placed in bags and pressed to extract the poisonous juice; it is afterwards baked on a hot iron plate in thin cakes like pancakes, of fifteen inches or more in diameter; these are kept constantly in motion to prevent burning.

BAMBOOS.

and removed from the fire as soon as they are crisp; when cool they are fit for use, and if kept dry will remain good for a long time. An intoxicating drink is prepared from the juice of mandioc by fermenting it with molasses; and tapioca is also made from the roots.

Besides those which have been mentioned, a multitude of valuable vegetable productions are found in South America, some of which are peculiar to that continent, and others common to it and other hot climates. Among these are coffee, the sugar-cane, maize, rice, plantains, bananas, the mango, the agave or pine-apple, olives, cotton, almonds, lemons, oranges, cinnamon, indigo, cinchona, many valuable timber-trees, dye-woods, spices, resins, and gums. Of these the banana is said to supply more nutritive food from a specific quantity of ground than any other known vegetable; while the mango is one of the most delicious fruits in the world.

It has already been remarked that the cactus is exclusively American. Those who are acquainted with them only from the specimens preserved in green-houses, would be surprised to see thick forests of these plants, thirty or forty feet in height, the trunks of some of which are four or five feet in circumference. Some are divided into upright branches like candelabra; some are of a globular form, enclosing a juicy pulp, which the mules break open the outer case with their feet to obtain, for the purpose of quenching their thirst in hot weather. Some of them have a diameter of two and a half or three feet, and weigh from 700 to 2000 lbs., while the Cactus nanus is so small, and so loosely rooted on the sand, that it gets between the toes of the dogs. To which may be added, that the wood becomes so hard from age that it resists both air and moisture for centuries; and the Indians, therefore, prefer it to any other for oars and doorposts.

Bamboos are especially East Indian, but some species are found in South America, which attain a gigantic size—from 40 to 60 feet in height—and are employed by the Indians as water-vessels, being cut just below a knot, and placed upright on the ground. They form one of the most beautiful ornaments of tropical vegetation. The form and disposition of the leaves give a character of lightness, which contrasts agreeably with their height. The smooth, glossy trunks, generally bend towards the banks of the stream near which they grow, and they wave with the slightest breath of air. The highest reeds of the south of Europe can give no idea of the gracefulness of their appearance. The knots are from ten to fifteen inches apart, and when green the hollow part is filled with clear water, so that about two quarts may be obtained from each division by making an incision in the cane; when they are nearly ripe, the water becomes like jelly; and when quite ripe it is converted into a white calcareous substance, two ounces being sometimes found in one of the divisions.

The ferns are not so common in tropical regions as many other forms of vegetation, but they sometimes occur of large size, and form one of the chief ornaments of the forests of the high lands. They sometimes become arborescent, rising to the height of twenty or thirty feet, the fronds spreading out in the form of a parasol, their delicate sprays extending widely on each side, and are some of the most remarkable of the vegetable kingdom—more beautiful even than the palms. Mountains and plains seem to be equally avoided by them; shady and damp situations, on a slightly undulating soil, with a moderately hot temperature, appear to be most favourable to their development.

The fibres of the leaves of several species of Bromelia, a kind of aloe, are formed into ropes, fishing-nets, and other articles. They vary from three to eight feet in length, and are covered with a kind of resinous substance, which preserves them from injury by water. A rope made of this material had been in constant use on the wharf of Paraiba for many years, when on one occasion it was required to hoist the heavy anchors of a line-of-battle ship on board a vessel. Ordinary ropes of larger diameter had been tried for the purpose, but were found insufficient, when this same old rope was employed and performed the task.

CHAPTER VI.

THE INDIANS.

AS animated beings cannot continue to exist without food, the population of a country depends in the first instance on the supply of food which it affords. Now, this has no natural tendency to increase: the productive powers of the soil, stimulated by heat and moisture, speedily put forth all their energies; growth and decay succeed each other in regular order, and without some interference the quantity produced in different seasons does not show much change. An increase in the population of a country without a corresponding increase in the supply of food will, therefore, reduce the proportion of each individual. Even the fruits of the palm-trees and other tropical productions are not sufficiently abundant to support a large population, and when from any casualty the supply falls short, no provision having been made, the inhabitants suffer and decline.

The first step in civilisation, then, is the adoption of some kind of culture of the ground for the purpose of obtaining food. The care of flocks and herds does not entail the necessity of settled habitations, nor does it require the same amount of care and foresight as agriculture; and those who are so engaged are little, if any, more civilized than those who live by hunting and fishing, as is shown by the state of the wandering Arabs and Tartars. But the cultivation of the soil involves the need of a settled residence, in order to protect and gather the crop, and promotes the habit of foresight by the necessity of making provision for the time which the crop requires before it comes to perfection. Wherever agriculture has been introduced the population has considerably increased. India is an example, as well as Mexico and Peru at the time of their discovery. The more rigorous climate of the temperate zones rendered greater industry in their cultivation necessary, and civilisation reached a higher point; the slower growth of the crops inducing a habit of accumulation, which is the foundation of all civilisation.

There are still many tribes in South America in an uncivilised, or partially civilised, condition; only a portion of its vast territory has been occupied by a population of European descent, and the remainder is overrun by tribes of wandering Indians. Varying, and somewhat contradictory characters, are given of these by different travellers. It appears probable that there are two originally distinct races, and the different branches of these races have received various degrees of civilisation, in proportion as they have had more or less intercourse with Europeans. All appear to be naturally indolent, and averse to regular occupation, and consequently disinclined to agriculture; but the tribes on the coast, and on the banks of some of the rivers, are somewhat more civilised than those in the interior. This is probably owing to the 'missions' which were established after the conquest; the priests who occupied them having exercised a beneficial influence over the natives by encouraging agriculture, and inducing them to settle in the neighbourhood of the missions, which were generally situated on the rivers. During the eighteen years which succeeded the expulsion of the Jesuits the farms of the missions were managed by Royal Commissioners, who were very negligent, killing the cattle for the sake of their hides, and allowing the cultivation to fall into decay, until there scarcely remained at the time of Humboldt's journey any trace of their former condition. 'All the actions of the Indian,' says Mr. Bates, ' show that the ruling desire is to be let alone; he is attached to his home, his quiet monotonous forest and river life; he likes to go to towns occasionally, to see the wonders introduced by the white men, but he has a great repugnance to living in the midst of the crowd; he prefers handicraft to field labour, and especially dislikes binding himself to regular labour for hire. He is shy and uneasy before strangers, but if they visit his abode he treats them well, for he has a rooted appreciation of the duty of hospitality: there is a pride about him, and being naturally formal and polite, he acts the host with great dignity. He withdraws from towns as soon as the stir of civilisation begins to make itself felt.'

The varying descriptions given by different travellers of the character and condition of the natives of South America may, therefore, easily be accounted for. Whether or not there were originally different races, the adoption under external influence of a different mode of living by different families, having so little intercourse with each other, is quite sufficient to have produced the differences now observable among them. They generally, according to Mr. Bates, present the same physical characteristics, modified by the circumstances and mode of life to which they are accustomed, and many of the moral characters in which they seem most to differ may be traced to the different development which circumstances have given to features which are common to all.

The Brazilian government have collected together some of the natives in villages called *aldeas*, where they employ themselves partially in agriculture, and also carry on some minor manufactures. M. Liais visited one of these, near Villa de Barriera, and found it to consist of about thirty dwellings constructed of palm-leaves, and forming a kind of street; the bad condition of the greater part of them manifesting the indolence of their owners. In passing through the village he saw most of the inhabitants, covered with rags, lying on bundles of reeds, or in

INDIAN VILLAGE.

hammocks. Some of them were sitting cross-legged before their doors, engaged in platting mats or baskets for sale at Barriera. This industry supplied their chief means of living, for they had only a few small inclosures planted with bananas, mandioc, or tobacco. Though the greater part of them were of pure race, yet many show signs of mixture with blacks and creoles. A shower of rain, which came on whilst M. Liais was passing through the village, opened the way for him to enter into conversation with them. One of them invited him to enter his hut. 'This proposal,' says M. Liais, 'which I at first took for politeness, was in fact instigated by curiosity. He wished to know what a stranger came to do in their village, and asked a multitude of questions on this point. His wife, seated on one of the bamboo benches which formed the only furniture of the hut, was occupied in swinging her infant in a small

P

hammock. Both of them, of an olive-brown complexion, presented the Guarani type without any mixture. This type, which has scarcely any of the Mongolian character, except the colour, and has neither the oblique eyes nor the prominent cheeks, is otherwise deficient in beauty. The half-closed eyes and thin lips greatly injure the expression of the face. Their hair is black and long; most of them wear it over the shoulders. The men are of low stature, and the beard scanty.

'When I was installed in the hut, several other Indians, men and women, entered. Their conversation was lively, for they are very loquacious. They related, respecting the subject of each question, a

INTERIOR OF INDIAN HUT.

thousand anecdotes about the people who had passed through the village, but they jumped continually from one subject to another, without their animation lessening, and superstition often mingled with their stories. On witnessing the want of connexion in their ideas, their suspicious nature, the spirit of stern independence which has remained notwithstanding their organization in villages, and also their indolence, I received the impression that it was impossible for them to have attained to a state of advanced civilisation but after the lapse of a long succession of ages; and that, notwithstanding the contact with the Caucasian race, they will not attain to it except by a fusion with European civilisation, similar to that of North America. Incapable of deep or prolonged reflection, the minds of these people receive only material images. At the same time they have received from Christianity the notions of good and evil, which seem to be altogether wanting among the fierce Botocudos.'

Mrs. Agassiz gives a more favourable representation of those on the Amazon. 'One of the Indians,' she says, 'invited us to continue our ramble to his home, which he said was not far beyond, in the forest. We readily complied, for the path he pointed out to us looked tempting in the extreme, leading into the depth of the wood. Under his guidance we continued for some distance, every now and then crossing one of the forest creeks on the logs. Seeing that I was rather timid, he cut for me a long pole, with the aid of which I felt quite brave, but at last we came to a place where the water was so deep that I could not touch bottom with my pole, and as the round log on which I was to cross was rather rocking and unsteady, I did not dare to advance. I told him in my imperfect Portuguese, that I was afraid: "*Naõ, mia branca, naõ tem medo*," ("No, my white, don't be afraid"), he said, reassuringly. Then, as if a thought struck him, he motioned me to wait, and, going a few steps up the creek, he unloosed his boat, brought it down to the spot where we stood, and put us across to the opposite shore. Just beyond was his pretty, picturesque home, where he showed me his children and introduced me to his wife. There is a natural courtesy about these people which is very attractive, and which Major Coutinho, who has lived among them a good deal, tells me is a general characteristic of the Amazonian Indians. When I took leave of them and returned to the canoe, I supposed our guide would simply put us across to the opposite shore, a distance of a few feet only, as he had done in coming. Instead of that, he headed the canoe up the creek, into the wood. I shall never forget that row, the more enchanting that it was so unexpected, through the narrow water-path, overarched by a solid roof of verdure, and black with shadows; and yet it was not gloomy, for outside the sun was setting in crimson and gold, and its last beams struck in under the boughs, and lit the interior of the forest with a warm glow. Nor shall I easily forget the face of our Indian friend who had welcomed us so warmly to his home, and who evidently enjoyed our exclamations of delight, and the effect of the surprise he had given us.'

Mr. Bates gives an equally favourable account on occasion of a visit which he paid to an Indian chief, about twenty miles from Ega. Having travelled for about three hours by the water-path, through an arched colonnade of trees, they arrived at the old man's dwelling in the heart of the virgin forest. The grounds were in neater order than in any sitios, even of civilised people, that Mr. Bates had seen on the Upper Amazons; and the evidences of regular industry and plenty were more numerous than usually found in the farms of civilised Indians and whites. Yet this man was a primitive Indian, who had never had much intercourse with the whites. They were received with courtesy, and the host apologised at dinner for not having knives and forks, waiting till all were served before he commenced; the women, as usual, not eating till after the men had finished. When the meal was over one of the women brought them water to wash their hands, in a painted clay basin of native manufacture, and a coarse, but clean, cotton napkin. On their return home in the afternoon they were loaded with presents by their generous hosts. These Indians belonged to the Passés, a tribe which is considered to be the most advanced of any in the district. Their industry, docility, and other good qualities, have from the first given them favour with the Portuguese, and Mr. Bates says he never heard of any violence having been committed, either by them or the settlers, in all their dealings together.

Another tribe of whom Mr. Bates gives a favourable representation are the Mundurucus, the most warlike as well as one of the most settled and industrious of the Brazilian tribes. They cultivate mandioca, and annually sell to the traders from three to five thousand baskets, of 60 lbs. each; they also collect in the forests salsaparilla, india-rubber, and tonquin beans. They have given up their old practice of cutting off and preserving the heads of their vanquished enemies, have adopted agricultural habits, are faithful to their treaties, and gentle in their manners, but show no liking for town life, and do not appear capable of any further progress in civilisation. There are some tribes which show more or less signs of civilisation, but none which have attained a very high state; on the other hand, there are tribes on which it seems to have produced scarcely any effect, but who have retreated, as it progressed, into the depth of the forests, retaining almost all their savage habits. Of these the Botocudos are one of the most irreclaimable.

The Botocudos inhabit the forests in the provinces of Minas Geraes and Espiritu Santo, and in physical appearance greatly resemble the tribes already alluded to, but are generally more robust. They move from place to place every day, traversing the thick forests in a state of absolute nudity, regardless of the strong and sharp thorns with which they abound.

They devour their food raw, or partially broiled on the fire; and after their meal throw themselves on the ground like a drove of swine, one serving as a pillow to his neighbour. The only industry which they exercise is the making of bows and arrows, with which they make war or obtain game for food; and of collars of the teeth of cabiai and jaguars, which they hang round their necks.

The Botocudos have received that name in consequence of their practice of inserting large plugs of wood in incisions made in their lips and ears;

BOTOCUDOS SLEEPING.

botoque being the Portuguese name of the bung of a barrel. They dislike the name, and call themselves *Enge ree moung*. Some of the tribes live quietly on the banks of the Rio Grande de Belmonte, others are always moving from place to place feared by all, being regarded as cannibals. When treated kindly they often show fidelity and attachment, and do not easily forget good treatment. About eight miles from Belmonte there lived a family who often received a young Botocudo, and treated him in a friendly manner. His compatriots sometimes made incursions in the neighbourhood. One day he ran to the house, breathless, and made signs that they must escape, for his tribe was approaching. No attention was paid to his warning, and a party shortly afterwards came and murdered nearly all the inhabitants.

They are very resentful. A mulatto soldier went one day to hunt with some Botocudos, in the forests near Belmonte. They were all peaceably disposed, but one of them having requested the mulatto to lend him his knife, and being refused, attempted to take it by force. The soldier, having made a motion as if he would strike the Botocudo, was instantly killed with an arrow.

They are very robust and capable of great endurance. Prince Maximilian von Weid, who has given the fullest account of them, says, 'Our people always returned quite exhausted from every excursion with the Botocudos. Their muscular strength enables them to go very swiftly in the hottest weather, both up and down hill; they penetrate the thickest and most entangled forests; they wade and swim through every river, if it be not too rapid; perfectly naked, therefore not incommoded by clothing, never getting into perspiration, carrying only their bows and arrows in their hand, they stoop with facility; and with their hardened skin, which fears neither thorns nor other injury, they creep through the smallest gap in the bushes, and can thus pass over a great extent of ground in a day. My hunters had experience of this their bodily superiority, among others, from a young Botocudo, named Jukeräcke: he had learned to be a very good marksman with his gun, and was at the same time uncommonly skilful in the use of the bow. I sometimes sent him with other Botocudos into the wood to kill animals; for a little flour and brandy they willingly hunted a whole day: Jukeräcke in particular was very serviceable, as he was agile, and showed much aptness to all bodily exercises. At first my hunters accompanied these people, but they soon complained that the Botocudos were too swift of foot, and let them hunt alone.'

The Botocudos are always dangerous companions; they are restrained by no law, internal or external; a slight accident may awaken their hostility, as in the case of the mulatto soldier; and they are very deceitful. The prince relates an instance of this. He had purchased a bow and arrows of an old man, who afterwards borrowed them on the pretext that he could not hunt without them. He asked the savage for them repeatedly, but in vain, and at last learnt that he had hidden them in the forest, and it was with much difficulty that he recovered them.

Prince Maximilian describes a singular scene of which he was a witness. A dispute had arisen between the chiefs of two parties of Botocudos, and 'one Sunday morning, when the weather was most beautifully serene, we saw all the Botocudos of the Quartel, some with their faces painted black, and others red, suddenly break up and wade through the river to the north bank, all with bundles of poles on their shoulders. Soon afterwards Captain June, with his people, came out of the woods, where a number of women and children had sought refuge in some large huts. Scarcely had the news of the approaching combat become known in the Quartel, when a crowd of spectators, among whom were the soldiers, an ecclesiastic from Minas, and several strangers, whom I also joined, hastened over to the field of battle. Each took for his security a pistol or a knife under his coat, in case the combat should be turned against us.

'When we landed on the opposite bank we found all the savages standing close together, and formed a half-circle about them. The combat was just beginning. First, the warriors of both parties uttered short rough tones of defiance to each other, walked sullenly round one another like angry dogs, at the same time making ready their poles. Captain Jeparack then came forward, walked about between the men, looked gloomily and directly before him with wide, staring eyes, and sang, with a tremulous voice, a long song, which probably described the affront he had received. In this manner the adverse parties became more and more inflamed; suddenly two of them advanced, and pushed one another with the arm on the breast, so that they staggered back, and then began to ply their poles. One first struck with all his might at the other, regardless where the blow fell. His antagonist bore the first attack seriously and calmly, without changing countenance; he then took his turn, and thus they belaboured each other with severe blows, the marks of which long remained visible in the large wheals on their naked bodies. As there were on the poles many sharp stumps of branches which had not been cut off, the effect of the blows was not always confined to bruises, but the blood flowed from the heads of many of the combatants. When two of them had thus thrashed each other handsomely, two more came forward, and several pair were often seen engaged at once; but they never laid hands on one another. When these combats had continued for some time they again walked about with a serious look, uttering tones of defiance, till heroic enthusiasm again seized them and set their poles in motion.

'Meanwhile, the women also fought valiantly amidst continual weeping and howling: they seized each other by the hair, struck with their fists, scratched with their nails, tore the plugs of wood out of each other's ears and lips, and scattered them on the field of battle as trophies.

If one threw her adversary down, a third, who stood behind, seized her by the legs and threw her down likewise, and then they pulled each other about on the ground. The men did not degrade themselves so far as to strike the women of the opposite party, but only pushed them with the ends of their poles or kicked them on the side, so that they rolled over and over. The lamentations and howlings of the women and children likewise resounded from the neighbouring huts, and heightened the effect of this most singular scene.

'In this manner the combat continued for about an hour, when all appeared weary. Some of the savages showed their courage and perseverance by walking about among the others uttering their tones of defiance. Captain Jeparack, as the principal person of the offended party, held out to the last; all seemed fatigued and exhausted, when he, not yet disposed to make peace, continued to sing his tremulous song, and encouraged the people to renew the combat, till we went up to him, clapped him on the shoulder, and told him that he was a valiant warrior, but that it was now time to make peace: upon which he at length suddenly quitted the field, and went over to the Quartel. Captain June had not shown so much energy; being an old man, he had taken no part in the combat, but constantly remained in the background.

'All of us had quitted the field of battle, which was covered with broken staves and ear-plugs, and returned to the Quartel. There we found our old friends Jukeräcke, Aho, Medeann, and others, covered with painful bruises; but their countenances showed to what degree man can harden himself against pain, for none of them seemed to pay the least attention to his wounds. They immediately sat down on their bruises, partly open, and ate the farinha which the commandant gave them.

'During the conflict the bows and arrows of the savages remained resting against the neighbouring trees, and no one had touched them; but it is said that on such occasions they sometimes have recourse to their arms, for which reason the Portuguese do not much like these combats to take place near the settlement. I did not learn till afterwards the cause of the fight which we had witnessed. Captain June and his people had hunted and killed peccaries in the grounds of Jeparack; the latter regarded this as a deep offence, for the Botocudos pay more or less respect to the limits of their respective hunting-grounds, and do not willingly trespass on them. Insults of this kind are the usual cause of their quarrels and wars. A short time after our departure a more serious contest took place on the return of Captain Gipakein, an ally of June.'

The prince also gives an account of a visit to another tribe, called Puris. These, as well as the Botocudos, are cannibals; the arms they employ are the same, *i.e.* a bow of six feet and a half in length, with arrows of six feet. Though living near a European settlement they were very scantily clothed, and many of them perfectly nude; painted with red and blue, with strings of hard berries, mixed with teeth of various animals, round their necks. On arriving at their encampment the prince's party found the whole horde lying on the grass. 'The group of naked brown figures presented a most singular spectacle. Men, women, and children were huddled together, and contemplated us with curious but timid looks. They had all adorned themselves as much as possible; only a few of the women wore a cloth round the waist, or over the breast; but most of them were without any covering. Some of the men had, by way of ornament, a piece of the skin of a monkey, of the kind called mono, fastened round their brows; and we observed also a few who had cut off their hair quite close. The women carried their little children partly in bandages made of bass, which were fastened over the right shoulder; others carried them on their backs, supported by broad bandages passing over the forehead. This is the manner in which they usually carry their baskets of provisions when they travel. Some of the men and girls were much painted: they had a red spot on the forehead and cheeks, and some of them red stripes on the face; others had black stripes lengthwise, and transverse strokes, with dots all over the body; and many of the little children were marked all over, like a leopard, with little black dots. This painting seems to be arbitrary, and regulated by their individual taste. Some of the girls wore a certain kind of ribbons round their heads; and the females in general fasten a bandage of bass or cord tightly round the wrists and ankles, in order, as they say, to make those parts small and elegant. Their huts are certainly some of the most simple in the world. The sleeping-net is suspended between two trunks of trees, to which, higher up, a pole is fastened transversely by means of a rope of bindweed, against which large palm-leaves are laid obliquely on the windward side; and these are lined below with heliconia leaves, and, when near the plantation, with those of the banana. Near a small fire on the ground lie some vessels of the fruit of the Crescentia cujete, or a few gourd-shells, a little wax, various trifles of dress or ornament, reeds for arrows, or arrow-heads, some feathers and provisions, such as bananas, and other fruit. The huts are small, and so exposed on every side that when the weather is unfavourable the brown inmates are seen seeking protection against it by crowding close round the fire, and cowering in the ashes; at other times the man lies stretched at ease in his hammock while the woman attends the fire, and broils meat, which is stuck on a pointed stick.'

When they returned to the plantation they presented the Indians with a small hog, which they obtained from their host. 'The hog was eating near the house, a Puri advanced softly and shot it too high, under the backbone; it ran away screaming, and dragging the arrow along with it. The savage then took a second arrow, shot the animal, while running, in the shoulder, and then caught it. Meantime the women had kindled a fire. When we all came up they shot the animal again in the neck to despatch it, and then in the breast. It was not, however, yet dead; it lay screaming and bleeding profusely: but, without regard to its cries, they threw it alive into the fire to singe off its hair, and laughed heartily at the groans which its sufferings extorted. It was not till our loudly-expressed displeasure at this barbarity became more and more impatient that one of them advanced and plunged a knife into the breast of the much-tortured animal, on which they scraped off the hair and immediately cut it up and divided it.'

This account exhibits the indifference with which they regard the sufferings of other creatures, as the former one does the indifference shown by the Botocudos to pain in their own persons. Insensibility and indolence are marked features in their character, with an unconquerable love of freedom and of a roving life. They have generally several wives, and upon the whole do not treat them ill, but they consider them as their property; they must do what the man commands, and are treated as beasts of burden, while he walks before with only his weapons in his hand.

The Puris, according to Mad. Pfeiffer, are distinguished for their skill in tracking runaway negroes. They smell the trace of the fugitive on the leaves of the trees; and unless the fugitive succeeds in reaching some running stream, in which he can wade or swim for a considerable distance, he seldom escapes the pursuit.

The Botocudos, the Puris, the Coropos, the Coroados, and other Indians of the eastern parts, use the large bow, and long arrows formed from the stem of a reed. They use three different heads for the arrows, according to the purpose for which it is to be employed. The first is used in war; the head is of broad cane, cut sharp at the edges and very pointed. The second is long, formed of airi wood, with many barbs on one side. The third has a blunt point, and is used to kill small animals. In some parts a peculiar kind of bow is used called *bodoc*. It is made of airi wood and has two strings, which are kept apart by two small pieces of wood; in the middle the two strings are united by a kind of net-work, in which a ball of clay or a small rounded stone is placed. The string and ball are drawn back together, and being suddenly let go, the ball is discharged. The boys begin exercising with this bow at a very early age, and attain great dexterity, killing small birds at a considerable distance, and even butterflies upon the flowers.

The tribes last named, the Coropos and the Coroados, are mostly settled, but had not, when the prince visited the country, altogether laid aside their

SETTLED INDIANS.

savage customs, having only recently shot one of the Puris. He found their houses good and roomy, constructed of wood and clay, the roofs covered with palm-leaves and reeds, their sleeping-nets hung up, and the bow and arrows resting against the wall in a corner of the room. The rest of their furniture consisted of pots, dishes, or bowls made of gourds and the calabash-tree, baskets of palm-leaves, &c. Their clothing was a white cotton shirt and breeches, but on Sundays they were better dressed, like the lower classes of Portuguese; but even then the men frequently went with their heads and feet bare. The women were fond of finery, and sometimes wear a veil.

The Indians, though naturally indolent, frequently manifest considerable quickness and intelligence. We have already referred to the case of a girl who was seized by an alligator, and saved herself by thrusting her fingers into his eyes; and Humboldt mentions another of a woman who, during an earthquake, when 35,000 Indians perished in a few minutes, saved herself and her children by extending her arms on each side. Mrs. Agassiz says, 'We were astonished at the aptitude they showed for the arts of civilisation so uncongenial to our North American Indians;' and Mr. Bates mentions a boy whom one of his acquaintances purchased of a trader, and employed as an apprentice at his trade of goldsmith, at which he made rapid progress, and after three months came to show Mr. B. a gold ring which he had made. Some of them are dirty in their habits, but others are remarkably cleanly, bathing several times a-day, and keeping their dwellings cleaner and more orderly than the European settlers. Some tribes are voracious, eating everything they can obtain, but will not touch strong liquor; others, on certain occasions, will drink as long as their liquor lasts: but they are generally temperate in their habits, and do not show that eagerness for strong drink so common among the natives of North America. They are curious but not rude. Mrs. Agassiz describes the women gathering about her and examining her dress, touching her rings and watch-chain, and evidently discussing her between themselves, but without roughness in their manner; and Mr. Bates, on his visit to the Munducucus, having produced some books of engravings to amuse the chief, shortly had a crowd of women and children about him. They would not allow him to miss a page, making him turn back when he tried to skip; but though there were fifty or sixty assembled, 'there was no pushing or rudeness, the grown-up women letting the young girls and children stand before them, and all behaved in the most quiet and orderly manner possible.'

Some of the Indians have considerable musical taste. Humboldt mentions one place where a fine church choir has been formed, consisting of natives, and a traveller had seen them playing the violin, the violoncello, the triangle, the guitar, and the flute.

They are very superstitious, and though many of them have been baptized, yet they seem to go to church chiefly for the sake of appearances, behaving quietly and decently, but without any real sense of religious truth, remaining motionless and apathetic; except at one time, when they made signs to one another that the priest was going to raise the chalice to his lips. Prince Maximilian mentions that in Minas Geraës there was an Indian who had become a priest, was generally esteemed, and resided several years in his parish, but all at once he was missed; and it was found that he had thrown off his clerical habit and run naked to his brethren in the woods, where he took several wives.

Occasions of rejoicing are celebrated by a feast with dancing and singing. Having prepared a large vessel full of *caouy*, a fermented liquor made from mandioc, maize, or batate, they smear themselves with long black lines, and some ornament their heads with feathers. One holds an instrument called *herenchedioca*, formed of the hoofs of the tapir fastened to strings, in two bunches, with which he marks the time; or sometimes they use another instrument, called *kekhickh;* which is a hollow calabash with a wooden handle, and containing some small flints. The dance begins by four men, who come forward, slightly leaning, and with measured steps describe a circle, following each other, and repeating with little variation of tone, '*Hoï, hoï, hé, hé, hé,*' the one with the instrument accompanying it, sometimes louder, sometimes fainter. The women then join the party in pairs, each with the left hand on the other's back; then men and women pass alternately round the jar which contains the liquor; continuing the dance in the middle of the day in the hottest season of the year, till the perspiration runs down. At intervals they dip a *coui* into the vessel and drink the *caouy*. They thus continue till the vessel is empty. Sometimes, after employing all the night in dancing, the young men go to the forest, and cut a large piece of a branch of barrigudo, which is very heavy, and driving a stick into each end to give a firm holding, one of them takes it on his shoulder and runs home; the others follow and try to take it from him, the contest continuing until they reach the place where the women are assembled, who give them their approbation. They then, though covered with sweat, plunge into the river to cool themselves; but it is said that some have died in consequence.

CHAPTER VII.

QUADRUPEDS, MONKEYS.

THE extraordinary variety presented by the animal kingdom in the tropical regions of the West has been already noticed. It has been asserted by naturalists that four hundred and eighty quadrupeds are peculiar to America; and of these a large number are to be found in the Southern part of the Continent. It must be acknowledged, however, that some of the nobler and more majestic animals are missing here, while, on the other hand, we meet with much that is peculiar and odd.

America, whether north or south, has neither lion nor tiger, neither leopard nor panther, neither elephant nor camel, although some of these have their representatives, or rather substitutes, in animals that resemble them only in some particular. On the other hand, these countries have their llamas, alpacas, and armadillos, besides a number of others, the flesh of which, though they may be wild animals, and sometimes most singular ones, yet is often found very tolerable eating — at any rate, by the natives.

Certain animals, too, South America has, in common with other parts of the world, though, perhaps, not very many; and even where this is the case, some marked peculiarity generally exists: as, for instance, in its species of the monkey tribe.

Certain other creatures are found of extraordinary dimensions, as the frogs, which are larger than elsewhere; also spiders, centipedes, and white ants; and in some parts insects exist in almost incredible numbers.

Boas are peculiarly American, and rattlesnakes are found nowhere else. Alligators are natives of South America; whilst many naturalists tell us that crocodiles are met with only in the Old World, others maintain that true crocodiles in South America are very numerous, and help to keep under the insect tribes. Vampires are known nowhere else.

As for the ornithology, it is rich and varied in the extreme. That of Brazil is the most extensive in the world; indeed, it is impossible to give any adequate idea of the magnificence of animal as well as vegetable life in these regions, though the distribution of both kinds is materially affected by the variety of character in the soil. Those vast forests which occupy the interior of the continent teem with almost innumerable specimens of the great Creator's handiwork, and afford endless proofs of His wisdom and might in the adaptation of the bodies of beasts, birds, reptiles, and insects, alike to the lives which they are designed to live and to enjoy.

The carnivorous animals in Brazil are comparatively few and small;

THE PUMA, OR COUGUAR.

yet, roaring through its forests, goes the couguar, or puma, which travellers have so commonly designated as 'the American lion.' His sole claim to this title consists, however, in his being a member of the feline species, possessing an unmarked skin of something the same colour as the lion's.

It is no compliment to the 'king of beasts' to make this one his namesake; for the puma has neither his mane nor his tufted tail, while as little does he possess the lion's courage. Moreover, his head is small; and he has nothing of the royal beast's natural dignity of character and majesty of mien.

He is, indeed, something of a coward; and on that account neither much feared nor much respected by the natives. He is found in North as well as South America; in fact, often seen in the United States. But he appears to prefer the south, and doubtless feels more at home in Brazil, Paraguay, and Guiana, because less disturbed, than in the more frequented north, where his range is constantly being contracted by the progress of civilisation.

The average length of the puma is about four feet, and his height two feet. He stands lower on the legs than the lion; and his head is round and small. In his natural state he is a sanguinary animal, killing far more of the smaller quadrupeds than he requires to eat, and seeming to take pleasure in torturing his prey. But as he is alarmed at the approach of men or dogs, and flees from them to the woods, he is not regarded as a very dangerous enemy. He climbs trees easily, and is the terror of monkeys. He roams through the upper regions of the forests, where he has undisputed hunting-ground, and fearlessly assails animals which cannot easily defend themselves, such as the horse, the mule, and the ass, and tears large pieces of flesh from their ribs; but he does not venture to meddle with oxen.

He even flies in the forest from the unarmed Indian, and it is only when severely wounded and driven into a corner that this animal commences a combat of despair; but then, becoming desperate, he sometimes kills the hunter.

The puma is easily tamed. D'Azara, the naturalist, had one which was as sensible to caresses as the common cat, and would play with a ball or orange like a kitten; and Mr. Kean, the tragedian, had a domesticated puma, which was much attached to him.

Lord Napier brought one home, and he described it thus:—'It is now about two years old,' he wrote. 'I have always observed it to rejoice in a large tub of water, jumping in and out, and rolling in the wet; very playful, with all the manner of the cat, without the treachery. It played with the dogs and monkeys, without even attempting to hurt them, or even returning an insult; but if an unfortunate goat or fowl came within its reach, it was snapped up immediately. It got adrift one night in London, and afterwards allowed a watchman to catch it in the streets without the slightest resistance.'

Although there have been instances of the puma attacking, and even destroying, the human species, in South America they have an instinctive dread of any encounter of this nature. Capt. Head, in his narrative of a journey across the Pampas, has the following anecdote of the puma, which, in common with other travellers, he incorrectly calls the lion:—

'The fear which all wild animals in America have of man is very singularly seen in the Pampas. I often rode towards the ostriches and pumas, crouching under the opposite side of my horse's neck; but I always found that, although they would allow my loose horse to approach them, they, even when young, ran from me, though little of my figure was visible; and, when one saw them all enjoying themselves in such full liberty, it was at first not pleasing to observe that one's appearance

COUGAR WATCHING DEER DRINKING.

was everywhere a signal to them that they should fly from their enemy Yet it is by this fear that man hath dominion over the beast of the field; and there is no animal in South America that does not acknowledge this instinctive feeling. As a singular proof of the above, and of the difference between the wild beasts of America and of the Old World, I will venture to relate a circumstance which a man sincerely assured me had happened to him in South America. He was trying to shoot some wild ducks, and, in order to approach them unperceived, he put the corner of his poncho (which is a sort of long narrow blanket) over his head, and crawling along the ground upon his hands and knees, the poncho not only covered his body, but trailed along the ground after him. As he was thus creeping by a large bush of reeds, he heard a loud, sudden, noise, between a bark and a roar; he felt something heavy strike his feet, and instantly jumping up he saw to his astonishment a large lion (puma) actually standing on his poncho; and perhaps the animal was equally astonished to find himself in the immediate presence of so athletic a man. The man told me he was unwilling to fire, as his gun was loaded with very small shot, and he therefore remained motionless, the lion standing on his poncho for many seconds; at last the creature turned his head, and walking very slowly away about ten yards, he stopped and turned again; the man still maintained his ground, upon which the lion tacitly acknowledged his supremacy, and walked off.'

The jaguar, so frequently called the American tiger, appears to be a more powerful and more daring animal than the puma. When well fed he sometimes attains the length of six feet. By some naturalists he has been confounded with the ounce of the Old World; yet he is properly to be distinguished alike from the tiger, the panther, the leopard, and the ounce. His special characteristics are these: he is of a bright fawn colour above; the body is marked along the spine with a chain of large eye-shaped rings, each of which has a black spot in the centre; along the sides are four chains of rings, but these are more oval than the others, and generally each contains two spots; the belly is white, with transverse black stripes. The face and sides of the neck are very thickly studded with black spots. The fur of the tail is not glossy; on the upper part the pattern is a zigzag, and not spotted like the body.

The hunters of South America affirm that there are two distinct varieties of the jaguar, the one being considerably smaller than the other, but both having certain marks in common. There are also black jaguars, which, however, are somewhat rare. The jaguar resembles the tiger in his ferocity and in the manner of seizing his prey. His usual haunts are in deep and retired forests, in the neighbourhood of some river. He is undaunted by the size of any animal; and he is not afraid of man, but springs on him, even though repelled by fire-arms, or by a blazing brand snatched from the midnight fire.

He seizes his victim with a sudden bound, breaks its neck, and then drags the carcase to some retired spot where he can enjoy it at leisure. D'Azara, in his account of Paraguay, mentions that he came upon a horse lying dead, and partially devoured, from which one of these animals had just been frightened. Conjecturing that the jaguar would return for his prey, he made his attendant drag it towards a tree, in which the traveller proposed to pass the night, so that he might be within shot of the animal if he made his appearance. This was accordingly done, but during his temporary absence, the animal returned, and although under the observation of the attendant, he dragged the horse in his teeth to the brink of a deep river in the neighbourhood, plunged in with his load, and having landed on the opposite side, retreated with the carcase into an adjoining thicket.

The jaguar causes such destruction amongst their herds and flocks, that the Indians often form hunting parties to go against him; and they have done much to reduce the numbers of these terrible animals, indeed in some parts they are almost exterminated They take with them a number of fierce dogs who drive the animals into the hollows of trees, where they are despatched by fire-arms, spears, or lances. Jaguars do not generally attack men by day, though they will spring on them at night; when once they have tasted human blood, however, they become more daring, or averse to other food; but when it is known that a jaguar has destroyed any person, the cause is made a common one; and the people in the neighbourhood join and pursue the enemy till they kill it.

The jaguar is not always victorious in his assaults on less formidable game, as is proved by the following narrative related by Stevenson:—
' During my stay at Esmereldas,' he says, 'I was requested to go into the woods, about a league and a half from the town, to see a perfect curiosity. Not being able to learn what it was, I went, and found the two hind-quarters of a full-grown jaguar suspended from the trunk of a tree, into which the claws were completely buried, all the fore-parts appearing to have been torn away, and fragments of it were scattered on the ground. This sight astonished me; and I was not less surprised at the account I received from the natives. The jaguar, for the purpose of killing the saino (a kind of pig) rushes on one of a herd, strikes it, and then betakes itself to a tree, which it ascends, and fastening its hind claws into the tree, hangs down sufficiently low to be able to strike the saino with its paws, which having effected in a moment, it draws itself up again to escape being hurt by the enemy. However, it appeared that in this case the jaguar had been incautious, and the saino had caught it by the paw, when the whole herd immediately attacked it, and tore as much of it to pieces as they could reach.'

A singular opinion is current among the inhabitants of the interior of Brazil, as to the preference of the jaguar for certain races of mankind,

'This fact, which I am happy to say I had no opportunity of verifying,' says one writer, 'was told me by the boatmen on the Rio de San Francesco one night, when, with a Brazilian engineer officer, we arrived at the junction of this river with the Rio Paraopebu. We were alone with four men, a soldier and four boatmen from the village of Morada Nova, whilst my wife and the rest of our party remained at that village, twelve miles in the rear, where we were to rejoin them to go on in another direction. When night arrived we proposed to land, and we perceived on the left bank a beaten path, formed by the wild animals in coming down to the river to drink, by which we could climb the bank, which was high and covered with wood. On approaching this landing place, we saw a herd of tapirs, who fled at our approach; and as we mounted the bank, a troop of monkeys, leaping about the trees, began to make a noisy song and took flight, after two shots from our revolvers. At the same time one of our men came to tell us that the ground was marked with traces of jaguars, and that we were near a place where these animals were accustomed to come to drink. But it was too late to seek for another place for our encampment, and we resolved to remain on the spot, and content ourselves with keeping up good fires. After having set up our tent, the men, of whom three were mulattoes, began talking, and I heard them speaking to the one of deepest colour, saying, "As for us, we have nothing to fear; if the jaguar comes it is thou whom he will choose." We then inquired the meaning of this conversation, and they told us that when Indians, negroes and whites were together, the animal always takes the Indian; when there is no Indian, he takes the negro; when no negro, the mulatto, and never takes a white man, except there is no one of any other race. However it may be, in the absence of Indians and negroes, the darkest of our mulattoes considered himself the most concerned to keep off the jaguars, and watched the fires the whole night, whilst the rest of our men and ourselves were in a profound sleep.' In the course of his travels, Humboldt had on one occasion an adventure with a jaguar on the banks of the Orinoco. As he was proceeding, his eyes directed towards the river, he discovered recent footmarks of a beast of prey, and turning towards the forest found himself within eighty steps of an enormously large jaguar. Although extremely frightened, he yet retained sufficient command of himself to follow the advice which the Indians had so often given, and continued to walk without moving his arms, making a large circuit towards the edge of the water. As the distance increased he accelerated his pace, and at length, judging it safe to look about, did so, and saw the animal in the same spot. Arriving at the boat out of breath, he related his adventures to the natives, who seemed to think it nothing extraordinary.

One day an Indian on his way to the mill saw a jaguar standing on

JAGUAR WATCHING AN ENCAMPMENT.

the road about ten paces in front of him as he emerged from the forest. The animal looked somewhat fiercely at him, and the Indian was puzzled for a moment, but summoning his presence of mind, he took off his broad-brimmed hat, and made a low bow, with ' Muito bem dias, meu senhor,' or 'A very good morning, sir.' Whereupon the jaguar turned slowly, and marched down the road with a dignified air.

Another traveller gives the following account: 'Three large animals of the cat kind are found in the interior of Brazil—the jaguar, the black jaguar, which is somewhat rare, and the couguar. The couguar is the least dangerous of these, though he attains a considerable size. It attacks only young beasts, whilst the jaguar of both breeds kill the largest cattle, and can carry them in their mouths for a considerable distance. They frequently kill several in one night, suck the blood, and devour the flesh at a later time. It is the custom to keep on a *fazenda* good dogs to hunt these dangerous animals; their track is followed, when, glutted with carnage, they have plunged into a thorny thicket to repose. As soon as the jaguar sees the dogs, he attempts to climb up a leaning tree; he is then fired on with guns to make him fall from this insecure refuge. But the chase is not always easy; the large jaguars do not so quietly yield to the dogs; they often kill one or two, carry them off, and devour them. There was near Valo a large jaguar, very famous, which never fled before the dogs. Three *vaquieros* having one day gone into the forest to seek their cattle, their dogs, prowling from side to side, found fresh traces of the jaguars, and followed them. The vaquieros had no guns, and were armed only with their long poles like lances, and deliberated whether they should take advantage of such an opportunity. They determined to do so, and boldly marched towards the animal, who stood with a threatening air in the midst of the dead dogs. The jaguar immediately attacked them, and wounded them one after another; but they struck him repeatedly with their poles, and gave him several wounds with the knife. One of them, whose courage failed, attempted to fly after having been wounded. The bravest was already stretched on the ground under the claws of the animal, when the timid took courage, attacked him with fresh vigour, and he was killed with blows of their poles. The men, grievously wounded, had much difficulty in returning home at night. They described the place where they had fought so bravely; a party went there, and found the jaguar lying in his own blood, and surrounded by several dead dogs. This adventure, generally known in that part of the country, and related by most trustworthy persons, proves that it is a mistake to accuse the jaguars of South America of cowardice. In the early times of its discovery, when ferocious animals were more numerous in the inhabited parts, there were many instances of men being attacked and killed by jaguars, though accidents of this kind were not so frequent as those which are related as having occurred in India and

Africa. Many authors—the Jesuit Eckart, for example—have related similar events.'

The Ocelot, called by the natives Le Chibigouazou, the Yagouarundi, or Eyra, the Margay, and Colocolo, are all animals of the wild-cat kind, indigenous in South America. The Ocelot is one of the most beautiful of its tribe; its fur is grey, slightly tinged with pale fawn, and the whole body is covered with shaded black stripes. It is timid and rather cowardly; it can rarely be tamed

TAPIRS.

by any caresses, and when wild shuns man and eludes dogs, by whom it is never taken. D'Azara, however, mentions one that was so tame that it was allowed perfect liberty, and became much attached to his master.

Dogs do not appear to have been among the aboriginal animals of this part of the world; but D'Azara found some specimens of a kind of wolf.

The Tapir, or Anta, called by the natives Mborebi, is one of the animals frequently attacked by the jaguar. This is a large animal, the flesh of which, though coarse, is eaten by the Indians. It most resembles the elephant of any American animal, measures 6 feet in length, and 3 feet 6 inches in height, has a long snout, and feeds chiefly on vegetable food, water-melons, wild fruits, and buds of young branches; but like the hog, which it resembles in its manner of drinking, it is not particular in its food, and has been known to gnaw very extraordinary things, as for instance a silver snuff-box full of snuff.

The tapir seems indeed a reckless kind of an animal, which rushes headlong after any object that it desires, seeking no path, but tearing desperately through or over every obstacle, for its skin being thicker than the bull's, it is not easily injured by anything with which it comes in contact. It goes alone, or in company with its mate: and the female is larger than the male. It is during the night that tapirs seek their food, which, as they are not beasts of prey, seems very singular. They swim well, and are perfectly indifferent as to the nature of their hiding-places during the day. Yet they are tenacious of life, though their safety lies in flight, not in any method of self-defence which they possess. There are two species, one of which is hairy; and they are easily tamed and domesticated if taken young.

Of all the South American quadrupeds the Tamandua, or ant-bear, is certainly one of the strangest, especially the largest species; for there are three, one being little bigger than a rat; another about the size of a fox; and a third, which measures from the tip of the snout to the end of the tail, something more than six feet. It is the largest kind, the great ant-bear, which is so very remarkable, both from its extraordinary appearance and character. Its head is small, and finishes in an extremely long, pointed snout; it has a broad black band on each side of the spine; and its tail is bushy, and of an enormous size.' This strange-looking creature, though so large and powerful, lives entirely on ants. He is perfectly inoffensive, and never injures men or property, or, indeed, any animal, except when attacked. He travels on through the recesses of the forest, choosing chiefly the low, swampy grounds, near creeks, ever in search of ant cities, and might live in peace to a good old age, were it not that his flesh happens to be very good eating; so that the Indian is ever in pursuit of him, as he is of his smaller game; and as he moves but slowly, he falls an easy prey to the poisoned arrow shot from a distance. For the Indians never attack him closely, knowing perfectly well the immense strength of his forelegs, and being fully aware that he possesses tremendous claws.

'These claws are his weapons of defence; and if attacked by any creature within reach he throws himself on his back, seizes his victim with these, and hugs him tight, till he expires from the pressure or is starved to death.

'Nor, if this process be ever so long, does the ant-bear suffer, as he can go longer without food than any animal, except the tortoise. Artists and

TAMANDUA, OR ANT BEAR.

many naturalists have formed an erroneous idea of the manner in which this creature walks, probably because so few specimens have been obtained, on account of the dread which the Indians have of approaching the animal. They have represented him as walking like a dog; but in reality, as soon as the ant-bear stands his feet take the form of a club hand. The claws are collected into one point, and drawn under the feet, while he walks entirely on the outer side of the foot. In this position he can walk without wearing down the points of his claws, which would not be the case if he went like a dog or cat, as he has not that retractile power which the feline species possess, by which they are able to preserve the sharpness of their claws even on the roughest paths.

'But why should so large an animal have such a small elongated snout-like head? It has a most strange and uncouth appearance in our eyes; yet we shall soon see that it is a wonderful instance of the adaptation of means to the attainment of a purpose. The ant-bear has no need of teeth, for he eats no food that requires chewing or gnawing. He has, therefore, no need for large jaws; and a wide mouth would prevent him from obtaining or disposing of the tiny insects on which he lives. What he wants is something that will give room for the reception and the working backwards and forwards of a long tongue, and further, something glutinous to moisten that tongue. And he has both; for inside the long snout which so well carries the tongue are two very large glands, situated below the root of it. From these is emitted a glutinous liquid, with which his tongue becomes lubricated when he puts it into the ants' nests. Thus admirably furnished, the ant-bear roams through the Brazilian campos, always in search of the wonderful cities of the white ant, which are dispersed over the plain in such incalculable numbers. Approaching one of them, he instantly strikes a hole through its circle of clay with his powerful hooked claws, and as the ants issue forth by thousands to resent the insult, he stretches out his tongue for their reception. They come out in furious multitudes, eager for revenge, immediately rush upon it, and vainly endeavouring to pierce its thick skin with their mandibles, remain sticking to the glutinous liquid of which I have spoken.

'The ant-bear waits quietly until his tongue is sufficiently loaded with its prey, then suddenly draws in his tongue and swallows them all.'

It is singular how this animal without teeth, without either the power to burrow or to fly swiftly from its foes, will yet range the forests in perfect security, fearing neither serpents nor jaguars, and relying entirely upon its powerful forelegs and claws.

'Richard Schomburgk,' we are told by a recent author, 'had an opportunity of seeing a young ant-bear make use of these formidable weapons. On the enemy's approach it assumed the defensive, and in such a manner as to make the boldest aggressor pause; for, resting on its left forefoot, it struck out so desperately with its right paw as would undoubtedly have torn off the flesh of any one that came in contact with its claws. Attacked from behind it wheeled with the rapidity of lightning, and on being assailed from several quarters at once threw itself on its back, and desperately fighting with its forelegs, uttered at the same time an angry growl of defiance.'

The ant-bear, though peaceably disposed, is indeed no coward. Even in a combat with the jaguar, which is considered the lord of the American forests, he appeared to be quite as often the victor as the vanquished; and that savage beast is not unfrequently found wallowing in its own blood after one of these encounters.

Yet the Indians have a way of killing it. It is useless to attack the creature with dogs; they can make no impression on his tough skin; and besides, his hinder parts are protected by thick and shaggy hair: but they have a poison into which they dip their arrows, and shooting with these they soon paralyse his muscles, and stretch him dead on the ground. Besides, they take advantage of a habit it has of turning its tail over its back, and standing still during rain; and when they meet with one in the woods they shake and rustle the trees. The animal mistakes the rustling for rain, and is quickly despatched by a blow on the head with a heavy stick.

This bushy tail is very useful to the curious creature. It serves him for tent as well as umbrella, so that he has no den, and seeks no other shelter. The first specimen of the Great Ant-bear was brought to the Zoological Gardens in the Regent's Park in 1853, at a cost of 200*l*. It came from the interior of Brazil.

The Skunk is an American animal much found in the north, but not confined to it. One species is found in Chili. It is a strange-looking animal with a black body, overlaid on the back with long white hair, which, however, leaves a broad black stripe down the back, while up the middle of the face is a white one. The creature's legs are not upright, so that it stands low, and its foreclaws are strong and very appropriate for digging. The great peculiarity of this quadruped is its power of emitting a yellow fluid of most abominable odour, and this seems to be its means of defence.

The Llama, which we may designate as the American camel, is a native of Peru. It is called by the Spaniards *carneros de la tierra*, and is found a very useful animal. The points of resemblance between the two animals are many, but there are also marked differences between them. The llama is smaller, but of a more elegant form than the camel. It measures from the sole of the hoof to the top of the head

4 feet 6 to 8 inches. The colour varies a good deal. Some are brown, with shades of yellow or black; some are speckled, and very rarely specimens are found quite white or quite black.

Llamas have no horns, but their small heads are adorned with a large tuft of hair. They have long slender necks, full round black eyes, short muzzles, and more or less cleft upper lips. Their bodies are handsomely turned, their legs long and slender, their feet bipartite, and the covering of their bodies a mixture of hair and wool, differing according to the species. For there are four kinds: the Llama, the Paco or Alpaca, the Huanaca or Guanaco, and the Vicuna or Vigonia. It is not in outward appearance that this animal resembles the camel. No; it is their internal conformation, and the adaptation of their bodies for a life in places where supplies of food and water are uncertain, that has obtained for them this appellation. Their stomachs are in a great measure similarly constituted; they are ruminating, and have four ventricles, the second of which is composed of two, and contains a number of cavities calculated for the deposit of water, so that they never show any wish to drink whilst they can obtain green pasture. The stomach is also provided with a store of nutritious matter, which, like the camel's hump, is absorbed as a compensation for occasional want of food; only, instead of forming a hump as in the camel, it lies in a thick bed of fat under the skin. The animals, however, instead of having to traverse deserts and endure great heat, are accustomed to climb the precipices of the Andes, and cannot bear extreme heat.

The camel's foot is uncloven, and has an elastic sole suited for the sand and stone over which it travels; whilst the two toes of the llama have strong nails, and each is furnished with a thick pad, enabling it to climb the precipices of the Andes, where neither ass nor mule could travel.

These animals have a callous covering over the breast-bone, on which they fall when reclining, either to sleep or to receive a burden; and this substance is also a defence against any contusion among the rocks. When sleeping they have their legs completely folded under them, and they rest on the breast. Their only means of defence is an ejection of viscous matter from the mouth.

Stevenson, from whose description many of the foregoing particulars are extracted, speaks of the llama as 'by far the handsomest and most majestic animal of the four species;' and he adds, 'in its portly appearance it is somewhat like a stag, but the gracefulness of its swan-like neck, its small head and mild countenance, add much to its beauty.' And he says that nothing can exceed the beauty of a drove of these animals as they file along the sides of the mountains, each following the other in the most orderly manner, and none ever needing either whip or spur to urge them on. No sight can be considered more interesting or characteristic of the country than such a drove of these useful animals.

Like the camel, they kneel to receive their burdens, which are of about a hundred pounds weight; but if too heavily laden they will refuse to run, and show their displeasure by spitting in the face of their oppressor.

On these journeys the first animal is decorated with a tastefully ornamented halter on its neck, covered with small hawks' bills, and a small streamer on his head. Thus they traverse the snowy tops of the Cordilleras.

The Alpaca of Peru is also used as a beast of burden; it is stronger than the llama, but more like a sheep in appearance, and its hair is much longer and softer. Its head is rounder, its legs shorter, and its body more plump. It will not follow a leader, but requires to be led by a string passed through a small hole in the ear. The two breeds dislike each other, and will not mix.

The Guanaco differs in shape and some other things from the former kinds. Its back is straight instead of arched; it is shorter in front than in its hind-quarters, and consequently, when pursued, rushes down, instead of up the mountains. But this kind likes the warm climates better than either of the others, and in winter will come into the valleys of its own accord.

The Vicuna is the smallest of the four species, being about the size of a goat; it is covered with very fine soft wool, of a pale brown colour, which makes good cloth.

The two last species are seldom domesticated; while the llama is now never found in a wild state, and the paco very seldom. They all four like to feed on the ichu that grows at the height of 14,000 feet above the sea, even in 18° south latitude. Young llamas are left with the dams for about a year, and then they are placed with the flocks. Most of these flocks are reared in the southern parts, and afterwards sent to the silver-mines of North Peru. The price of a strong, full-grown llama, is from three to four dollars, but sometimes one may be had for a dollar and a half. The introduction of horses, mules, and sheep, has lowered their value, otherwise the price was much higher soon after the conquest of Peru.

They are of the greatest use in the transport of silver from the mines, for these are sometimes at a great elevation, and the metal has to be brought down such steep declivities that neither asses nor mules could keep their footing.

The Indians also often take large flocks of them to the coast to procure salt; but as the animals will not feed at night, the journeys have to be short to allow time for grazing. Llamas, when resting, make a strange humming noise, which at a distance, proceeding from a whole flock, sounds very much like a number of Æolian harps.

If when travelling any strange noise is heard, and startles them, the flock immediately disperses; and their drivers, the arrieros, have great difficulty in reassembling them.

The Indians are as fond of these creatures as the Arabs are of their camels and horses. They love to decorate them with ribbons, and often affectionately caress and fondle them. 'If,' says Tschudi, 'during a journey one of these animals is tired and lies down, the arriero kneels beside the creature, and addresses to it the most coaxing and endearing expressions. But, notwithstanding all the care and attention bestowed on them, many llamas perish on every journey to the coast, as they are not able to bear the warm climate.'

It is not true, as some old writers have asserted, that Indians use these creatures for riding and for draught. An Indian may occasionally, in crossing a river, mount on the back of one, but he will certainly dismount on reaching the other side.

Their flesh is spongy, and not agreeable in flavour; the wool is used in making coarse cloths.

That of the alpaco, on the contrary, is not only made into blankets, &c., but is exported to Europe, where it is largely used. These animals are kept in large flocks, and graze all the year on the level heights. They are remarkably obstinate, and very shy. When one of them is separated from the flock he throws himself on the ground, and cannot be induced to rise. The huanaco is a larger animal, and more resembles the llama, but it is almost impossible to tame them. The vicuna is more beautiful than either of the two latter kinds, and its back and thighs are of a peculiar reddish yellow, called by the natives *color di vicuna*.

It never ventures up to the rocky heights, for its hoofs are only accustomed to soft, turfy ground. Whilst the females are quietly grazing the male stands some paces apart keeping guard, and at the approach of danger he makes a kind of whistling sound, and accompanies it by a quick motion of the foot. At this the females draw together and flee, while the male covers their retreat, and frequently stops to watch the movements of the enemy. All this watchful care is returned by the most singular fidelity and affection on the part of the females, and if he is wounded or killed they rally round their protector, and will suffer anything rather than desert him.

The Indians catch them by means of what they call *chacu*, large parties of men joining in the hunt, and some women accompanying them to cook for the hunters. Seventy or eighty persons thus go to the Altos, which are the haunts of the animals, taking with them stakes and a quantity of rope and cord. A spacious plain is selected, and the stakes are driven into the ground in a circle at intervals, and these are connected by ropes at about two feet from the ground. So a space of about half a league is enclosed, with a wide entrance. Coloured pieces of rag are tied to the ropes by the women, and left to flutter about. Then the men range about on horseback, driving all the herds they meet within the circle, which is at last closed. The timid animals, frightened by the fluttering rags, do not attempt to leap over the ropes; and the Indians easily catch them by the bolas, which are three balls of stone or metal strung together, and then swung at the legs of the animals, round which they twine. Then they are killed, and the flesh divided among the hunters. When all the animals are killed, the stakes and ropes are taken away to another spot, and the same thing is done again. 'During five days,' says Tschudi, 'in which I took part in the chase, 122 vicunas were caught.'

Under the dominion of the Incas the Peruvians rendered almost divine worship to the llama and his relatives, which exclusively furnish them with wool for clothing, and flesh for food. Their temples were adorned with representations of them in gold and silver images; and they were carved on every kind of domestic utensil.

We will now turn our attention to a very singular specimen of the animal world, and one as different as can well be imagined from those just described. This is the Sloth, a much-maligned and misunderstood quadruped. It has been said over and over again that this creature is the very embodiment of misery; that it is always sighing and groaning, and appears to have no enjoyment whatever in its little life. The fact is, however, that the sloth as usually seen is in a position for which he was never intended. He is just another example of the adaptation of means to an end, always to be noticed in the works of God, but remarkable in this respect, that he cannot adapt himself, as most other creatures can, at least in some degree, to a life for which he was not fitted by nature. Birds, for instance, are formed to fly through the air, but they can be domesticated or made happy in captivity. The sloth, on the contrary, is made to live in trees, and in trees alone; and not only so, for it is not *on* the branches that he is comfortable and happy, but suspended to them; and in order to do this, as Waterton says, 'he must have a very different formation from that of any other known quadruped.'

The following is a description of this curious animal, furnished by Dr. Herrtado of Guayaquil:—'The snout short, forehead high, eyes black, almost covered with long black eye-lashes, no incisors in the upper jaw, four legs ill-formed, thighs ill-shaped and clumsy, hind-legs short and thick, the toes united, having three long curved claws on the hind and fore-feet, twenty-eight ribs, three stomachs, very short intestines, very short tail, and the whole length of the body between four and five feet.'

Put this animal on the ground, and he is the very picture of misery and awkwardness: it is covered with long, shaggy hair, resembling dried grass; its motion is very slow, and at each step it howls most hideously, and scarcely walks ten yards in as many hours. Thus it has obtained the name it bears, and likewise its melancholy character. But the fact is, that the animal having no soles to its feet really cannot walk, and only if the ground be rough and uneven can it drag itself along at all. In all probability it really is in pain in such a position, and this will account for its beseeching look and its miserable groans and sighs. Waterton says that he once had a good opportunity of watching one that was in the house with him for some time, and observed that he was all right as soon as he came in contact with the branch of a tree, and that he always made for one if he took him out. In truth, the sloth has no business on the ground at all, nor any motive for activity. He feeds on the leaves and buds of trees, and therefore a tree is his natural home. It is his habit to get to the topmost bough of one and there remain until he has eaten every leaf. Even for some time afterwards will he stay there, crying and howling until hunger obliges him to search for food. Then the creature forms itself into a round lump, and drops from the tree upon the ground as if devoid of life. When in a tree, however, the sloth does not hang head downwards like the vampire. When asleep he supports himself from a branch parallel to the earth. He first seizes the branch with one arm, and then with the other, and after that he brings up both his legs, one by one, to the same branch. Thus he seems quite at rest, and his short tail of an inch and a half does not either interfere with the tight clasp of legs and arms, nor does it hang down to become the prey of any animal or insect. In climbing he does not move both arms together, but puts out first one and then the other. His hair too is different from any other animal, for it is thick and coarse at the extremities and fine as a spider's thread at the root. It is so nearly the same colour as the moss that grows on the branches that the animal is hardly to be perceived when at rest.

One kind of sloth has a bar of fine black hair on his back, and on each side of it another bar of yellow. His fore-legs also are strong and muscular, exactly suited to enable him to climb and hang as he does. He lives in the thickest forests away from the haunts of men, and is most active when the wind rises and the thick boughs are interlaced together. No one who saw him then passing from one tree to another would call him Sloth.

'One day,' says a writer who has been before quoted, 'as we were crossing the Essequibo, I saw a large two-toed sloth on the ground, and upon its back: how he had got there nobody could tell, the Indian said he had never surprised a sloth in such a situation before. He would hardly have come there to drink, for both above and below the place the branches of the trees touched the water, and afforded him a safe and easy access to it. Be this as it may, he could not make his way through the sand in time enough to make his escape before we landed. As soon as we got up to him he threw himself upon his back, and defended himself in gallant style with his fore-legs. "Come, poor fellow," said I to him, "if thou hast got into a hobble to-day, thou shalt not suffer for it; the forest is large enough both for thee and for me to roam in: go thy way up above, and enjoy thyself in these endless wilds. It is more than probable thou wilt never have another interview with man, so fare thee well!" On saying this, I took up a long stick which was lying there, held it for him to hook on, and then conveyed him to a high and stately mora. He ascended with wonderful rapidity, and in about a minute he was almost at the top of the tree. He now went off in a side direction, and caught hold of the branches of a neighbouring tree; he then proceeded towards the heart of the forest. I stood looking on in amazement at his singular mode of progress. I followed him with my eyes till the intervening branches closed between us, and then I lost sight for ever of the two-toed sloth. I was going to add that I never saw a sloth take to his heels in such earnest; but the expression will not do, for the sloth has no heels. Indians like the flesh of the sloth, and say it is very savoury.'

Besides the two-toed sloth, or unau, there is the three-toed, or ai, so called from its feeble, plaintive cry, and the collared sloth, with some different characteristics. The hair of the unau and ai is long and withered-looking, that of the collared sloth coarser and more frizzled.

We turn now to another quadruped whose mouth, like the great ant-bear, and the sloth's, is snout-shaped, viz. the Armadillo; a creature which, instead of hair, is covered with a kind of armour composed of three hard, bony bucklers, on the head, shoulders, and rump respectively, the two last of which hang down and protect the belly. These bucklers appear to be composed of a series of bands, connected however by transverse bands, which make the armour pliable.

There are several species of armadillos, but they are confined to the warmer parts of America, and none of them attain any considerable size. One, which is called the mulita, or little mule, on account of its long ears, is about eight inches in length; another, sometimes called bolo, is about eighteen inches long to the tip of the tail. These creatures feed partly on vegetables; but they also eat eggs and young birds when they find them on the ground, and greedily devour worms, lizards, frogs, and even vipers.

But there is another kind of food which they love better still. All over the plains of South America are to be found the carcases of wild cattle, which have been killed for their skins and tallow alone. All kinds of carnivorous animals flock to these, among them many an armadillo. They are, therefore, among the scavengers of the deserts and of the fertile plains. They run very fast, though always in a straight line, as their armour only allows them to turn in a circular manner, and are quickly attracted by the smell of the putrid carcase. The armadillo is a burrowing animal. The female brings forth her young, three or four every month, in a hole that she has made for herself, and feeds them on fruits and vegetables. When pursued, if on the mountains, they roll themselves up, and fall down precipices to elude their pursuers, but on the plains they are easily caught. The armadillo roasted in his shell is thought a great delicacy by the Spaniards and Portuguese; but the natives separate the shells, clean the meat, and season it with capsicum, salt, onions, and herbs, place it one shell, and covering it with the other, stew it in an oven.

Another snout-faced animal is the Peccary, which is a kind of pig indigenous in South America, and indeed forms one of the only two

TATOU, OR ARMADILLO.

quadrupeds of that class which belong to the fauna of that continent, the other being a species of Babirousa,—literally, hog-deer.

There are two kinds of peccary, the white-lipped and the collared. The former congregate in enormous herds, and, guided by one which acts as leader—so the Indians say—traverse large districts and cross the most rapid rivers in their search for food. They will often devastate plantations, digging in the ground for roots, potatoes, maize, or any kind of food which they desire. They are often attacked by huntsmen, and are not easily put to flight. Indeed they become enraged by any attack, chatter with their teeth, grunt, and, if able, attack and tear their enemy to pieces. But they cannot climb; and solitary travellers thus endangered, if wise, take refuge in a tree should one be within reach.

The collared peccary, so called from a yellowish-white stripe on the shoulders, goes either in pairs or small parties, and is more timid than the other. It lives generally in the hollows of the earth, or in trees. It is about three feet long. The flesh of both animals is eaten. The common hog has now, however, become a denizen of the woods, and lives in a wild state, though it was at first introduced from Europe.

CABIAI.

The Cavies, or agoutis, are strange-looking creatures of the rodent genus, which are very numerous in South America. Their great peculiarity is in their feet and legs, which strongly resemble the claws of birds. They have no power to climb, on account of the stiffness of these claws. Their hind-legs are much larger than the fore ones, and they use their fore-feet as paws, sitting, when eating, on their hindquarters. They have all small ears rounded at the tips. One kind, called 'The long-nosed agouti,' is very common in Brazil and Guiana, although it also inhabits the West India Islands. This animal seldom burrows; it generally inhabits the hollows of trees, and runs like a hare, with long leaps. Its flesh is considered very good, by colonists as well as by Indians. Another species of cavy is the Cabiai, which is described as a kind of Guinea-pig without a tail.

The agouti must be very prolific, for it is still numerous, although preyed on by animals as well as man. Brazil has likewise its porcupine, but it differs very much from its brethren in other climes; the snout and bristles being short, and the tail very long, and prehensile, too. This animal is a very good climber. Other porcupines also are found resembling those of Africa. There is one in Brazil which in form very much resembles a mouse; in this genus the spines are small, and not conspicuous. There is another species of a reddish colour and more spiny, which inhabits Guiana.

One species of an allied family pierce and excavate the ground for hundreds of miles in the same neighbourhood, so as to make riding very dangerous, for the horses continually catch their feet in the holes made by these Biscacha or Viscacha, for the name seems to be spelt either way. The little animal is very clever in the construction of its dwelling, and often large families make a settlement together, to which there are several entrances, all so contrived that no rain can get into the apartments below. Then they come out and sit at their respective openings in the evening; and if everything is quiet they go after dark and make devastation among the maize and wheat. They conceal the openings to their settlements by heaping up all sorts of odds and ends of stick, bones, &c. One great traveller has described them as the most serious-looking of quadrupeds, with mustachios and grey hair.

A small kind of owl is often seen at the entrances, and as they frequently run down the burrow it is supposed that they either share the dwelling or eject the biscachas from their homes by force.

These two last animals much resemble the chinchilla, a little creature very much like a squirrel, whose beautiful fur has given it importance. Ever since the conquest has the chinchilla been highly prized on account of its beautiful soft fur. It is a very clean little beast, and most sociable when tamed. When still it usually sits on its haunches, and it can stand upright on its hind-legs. When tamed, it is found very playful.

We turn now to a family of larger quadrupeds, the distinguishing mark of which is, that instead of walking on a part of the foot, the toes or the sides, they support themselves like the human species, on the whole sole of the foot. The Bear is the most familiar example of this class, and bears are common in North America, both the ordinary brown bear and his polar brother. Not so as regards South America. There the spectacled bear seems the representative of this branch of the family—a curious little animal, with a long, wise-looking face, and something about his eyes which gives him the appearance of wearing spectacles. He has smooth, shining black fur, a long nose, small buff-coloured jaws, and a line of the same colour over the eyes. And besides, he has a white throat. This little fellow inhabits the mountains of Chili.

The Racoon is another quadruped belonging to the same family to be found in these regions, though it goes also far north. The one found

COATIS, OR SOUTH AMERICAN RACOON.

in North America is rather larger than a badger, clumsy and awkward in its gait, and its fur being of two kinds seems to form for it a double covering. It has a head something like a fox, and a very acute smell. Like the bear its food is of all kinds. It rolls itself up and sleeps by day, and goes in search of its prey at night. Oysters are a favourite food with it, and it catches them in a curious way. Standing by the waterside the creature watches the oyster as it opens its shell, then adroitly puts in its paw and seizes its prey.

Those found in South America are long-nosed creatures, called by the French, Coatis; they differ from the North American genus in the longer muzzle and longer tail. There are two species, one red and the other brown. The noses of both are in perpetual motion, digging and discovering worms. In feeding, these coatis generally convey their food to the mouth by means of the sharp nails of their fore-paws. They live in companies, and emit a very disagreeable odour.

Pouch-bearing animals also are represented in this part of the globe, for South America has its opossum. Stevenson says that he found it in the valleys along the coast, and describes it as 'about two feet long, including the tail, which is as long as the body. The nose is pointed like that of a hog, and has no hair on it from the eyes to the mouth. The ears are thin, without any hair on them, and stand erect. The feet also are naked and small, and it holds its food with its fore-paws like a monkey. The body is covered with hair, black at the roots and white at the points, which gives it a shady grey colour. The tail is slender and naked, and by it the creature can hang suspended to the branch of a tree. The female brings forth four or five young ones at a time, not larger than mice when first born, and they immediately betake themselves to the pouch under the belly of their mother. The pouch is formed by a fold of the skin hanging on the outside, and covered with a very soft down, or fur, in the inside. The nipples are so situated that the young can reach them as they are carried about by their mother. When about the size of full-grown mice they leave the pouch by an opening in the centre, and bask in the sun, but if any danger threatens them they immediately take refuge in their natural home. I one day caught an old opossum by the tail, when four of her young ones ran out. I chased and captured two of them, and they immediately hid themselves by running up the inside of my coat-sleeves. I took them home, reared them, and they became perfectly domesticated, were very tame, and would sleep on the same mat with a dog. They feed on fruits or esculents, will eat flesh, and are particularly fond of eggs. The Indians esteem them as food, but I never had an opportunity of eating any. The natives sometimes call the opossum *mochilera*, from mochila, a knapsack; the Indians call it *muca muca*.'

But as the opossum is by no means peculiar to South America, so neither is a very different animal, to which it is now time to turn our attention. Yet if it has not a monopoly of the monkey tribe, South America is certainly well represented in that family, and in those same deep forests where the puma and jaguar roam, and through which the great ant-bear walks almost undisturbed, while the sloth passes from tree to tree, and the armadillo hurries after its food; there, above them all, and leaping with wild gambols through the branches, while their chattering cries and shouts ring through all the woods, are the many members of the monkey tribe, as noisy and happy as the day is long.

Now these Monkeys of the New World all differ in one respect from those of the Old, and that is, in the shape and lateral position of the nostrils. Look at the nose of some Eastern monkey, and you see that it differs but little from that of a human being, except in not being so prominent or imposing a feature; but take a South American specimen, and the two nostrils are almost like two separate noses joined at the top.

In the Old World again there are monkeys without any tails, or with very short ones. But all South American monkeys have long tails, naked underneath; tails also which naturally form a hook at the end, grasp anything tightly, and indeed have all the powers of fingers. Thus, like several other animals which we have noticed, these monkeys can spring from tree to tree, can cling to the boughs, and have their homes entirely in those many acres of boughs. Indeed, American monkeys are awkward on the ground, and they require this sort of tail the more because their thumbs are almost universally imperfect. The howling monkeys are among the most noticeable. The peculiar construction of their throats, and their robust nature, give them a power of voice which is perfectly astonishing. Humboldt discovered a group of trees in which some of this family lived, and found that he could hear them in the daytime at a distance of 24,000 feet, and much farther off by night. The Indians, he said, insisted that they always had a leader in their song; and they likewise assured him that to drink out of the bony drum of one of them would cure any asthma. He calls them Araguatos, and says that they resemble young bears, with bushy, reddish-brown hair, blackish-blue faces, and bare, wrinkled skins, with long prehensile tails. He took their howling as indicative of a coming storm.

Then there are the Coaitas, or spider monkeys, so named from their long, slender limbs and very long tails; one of these, called the cayou, is quite black.

The brown coaita has all its hair turned back from its face, and is said to be as intelligent as a dog and as obedient as a well-trained one.

Again, the Capuchin is an American monkey, distinguished by its round skull and short muzzle; and Humboldt was particularly delighted with the titi monkey, one of which he purchased. He mentions a very interesting fact, illustrative of the sagacity of this creature. One day, having a work on natural history in his hand, he showed it to the monkey, when it instantly put out its hand to grasp a grasshopper or wasp; which was the more remarkable, as the plates were not coloured.

On another occasion he saw a monkey of a new species, which every day seized a pig in the court of a house, and remained upon the poor creature all day in its wanderings in the savannahs. He also met with a large creature of the ape kind, which was said to build huts and eat human flesh.

Then 'wing-handed' as well as 'four-handed' creatures abound in these forests; and vampires, together with other similar bats, are found nowhere else.

Buffon has described the vampire bat for us—a creature measuring from 26 inches to 32 inches from wing to wing. 'It has,' he says, 'a long muzzle, and the hideous aspect of the most ill-favoured bats. Its head is surmounted with huge ears, and there is a membrane above the nostrils in the form of a horn, or a pointed crest, which greatly augments the deformity of the face.' This creature inhabits hollow trees and deserted houses, and may be seen sometimes hanging head downwards in clusters from the branch of a tree.

Doubts have been thrown on the stories of the vampire's bloodthirsty disposition by some naturalists, who insist upon it that *all* bats are insectivorous only. But Mr. Darwin was present when one was actually caught upon a horse. His servant, fancying that the creature was unusually restless, went to see what was the matter, and thinking he saw something on his shoulder put out his hand and caught the vampire. The place where he had bitten the horse was slightly swollen and bloody. In some parts the vampire is called 'the doctor.'

Mr. Waterton does not agree to the opinion that vampires live wholly on the blood which they suck from the veins of their victims. He lived at one time in a deserted house, in and out of which one of these creatures continued to fly, and frequently saw him making a meal on some juicy fruit.

Still there is no doubt that their habit is to go about at night, and prey in this manner on any defenceless animal that they can find. But they do not always kill their victims at one meal. Sometimes they return again and again, and have to be kept off by some kind of ointment, of which they dislike the smell.

CHAPTER VIII.

BIRDS.

NOW, leaving the four-footed race, let us turn our attention to the feathered tribes of South America, and inquire what specimens of the many-winged and many-voiced tribes there exist, making those sunny lands more beauteous, and causing forest and meadow, rock and mountain, to resound with song.

We begin with the birds of prey, well represented in the gigantic Condor. Only within the last half century has he been known in this country, for his character was long both misunderstood and misrepresented, so that it was thought until recently quite out of the question to bring any of his formidable family home to be studied and admired. And yet the bird has attained to a marvellous celebrity, for 'it was known in Europe,' says Tschudi, 'at a period when its native land was numbered amongst those fabulous regions which are regarded as the scenes of imaginary wonders. The most extravagant accounts of the condor were then written and read, and general credence was given to every story which travellers brought from those far-distant lands. It was only at the end of the present century that Humboldt overthrew the extravagant notions that previously prevailed respecting the size, strength, and habits of this extraordinary bird.'

Like the carnivorous quadrupeds, all these birds of prey are fitted by their construction for the life that they are destined to live; and repulsive as the idea of that life is to many a mind, perplexing as is the existence of such creatures to many another, yet let us consider how this earth would get on without some such beings. What kind of existence would man lead were there no check to the multiplication of the many noisome creatures and insects which afford him trouble and annoyance, even as things are?

Yes, such beings are necessary in the present state of things, and therefore have the vultures, and eagles, and condors, those natural weapons which enable them to make their power felt,—the strong, crooked beaks, trenchant and sharp at the edges, acute and curved at the points; the four toes so strongly armed with powerful talons, long, curved, and pointed; the broad breast-bone without any opening, affording so firm an attachment for the muscles of the large, sometimes enormous wings, which are requisite for their support in their long and lofty flights.

This bird is very closely allied to the vultures of the old world, differing from them chiefly in the large, fleshy and rather cartilaginous caruncle which surmounts its beak; in the large size of its nostrils, placed almost at the very extremity of the cere, or naked part at the base of the beak; and in the comparative length of its quill-feathers, the third being the longest of the series.

'The full-grown condor,' says the author whom we quoted above, 'measures, from the point of the beak to the end of the tail, from four feet ten inches to five feet, and from the tip of one wing to the tip of the other from twelve to thirteen feet. This bird feeds chiefly on carrion; it is only when impelled by hunger that he seizes living animals, and even then only the small and defenceless, such as the young of sheep, vicunas, and llamas. He cannot raise great weights with his feet, which, however, he uses to aid the power of his beak. The principal strength of the condor lies in his neck and in his feet, yet he cannot when flying carry a weight exceeding eight or ten pounds. All accounts of sheep and calves being carried off by condors are mere exaggerations. This bird passes a great part of the day in sleep, and hovers in quest of prey, chiefly in the morning and evening. Whilst soaring at a height above the reach of human eyes, the sharp-sighted condor discerns his prey on the heights beneath him, and darts down upon it with the swiftness of lightning.'

The general colour of the bird is a glossy black; but part of the wing in the male is white; and it has a white ring of soft feathers round the neck. It has also a white beak, or rather, mostly white; while the head and neck are of a reddish purple.

The condor goes in pairs, or in small numbers, not in large flocks like the vulture; only when some beast is slain, and the carcase left, then, down from the peaks in the Cordilleras, from their haunts at the height of 10,000, or 15,000 feet above the sea, down come these giant birds in troops, whether informed by sight or smell is uncertain, though it is known that they possess both senses in great perfection. Mr. Darwin thus describes their motion: 'When the condors in a flock are

THE CONDOR.

wheeling round and round any spot their flight is beautiful. Except when rising from the ground I do not recollect ever to have seen one of these birds flap his wings ... They moved in large curves, sweeping in circles, descending and ascending without once flapping. As they glided over my head, I intently watched from an oblique position the outlines of the separate and terminal feathers of the wing; if there had been the least vibratory movement these would have blended together, but they were seen distinct against the blue sky. The head and the neck were moved frequently, and apparently with force; and it appeared as if the extended wings formed the fulcrum on which the movements of the neck, body, and tail, acted. . . . It is truly wonderful to see so great a bird, hour after hour, without any apparent exertion, wheeling and gliding over mountain and river.'

' Some old writers have affirmed,' says Tschudi, ' that the plumage of this bird is invulnerable to a musket-ball;' which is, of course, absurd. However, it is so thick and strong, that,' he continues, 'the natives seldom attempt to shoot the condor; they usually catch him by traps, or by the lasso, or kill him by stones flung from slings, or by the bolas. A curious method of capturing him alive is practised in the province of Abancay. A fresh cowhide, with some fragments of flesh adhering to it, is spread out on one of the level heights, and an Indian, provided with ropes, creeps beneath it; some others station themselves in ambush near the spot, ready to assist him. Presently a condor, attracted by the smell of the flesh, darts down upon the cowhide; and then the Indian, who is concealed under it, seizes the bird by the legs, and binds them fast in the skin, as if in a bag. The captured condor flaps his wings, and makes ineffectual attempts to fly; but he is speedily secured, and carried in triumph to the nearest village.

'The Indians quote numerous instances of young children having been attacked by condors. That the birds are sometimes extremely fierce is very certain. The following occurrence came within my own knowledge whilst here. I had a condor, which, when he first came into my possession, was very young. To prevent his escape, as soon as he was able to fly he was fastened by the leg to a chain, to which was attached a piece of iron of about six pounds weight. He had a large court to range in; and he dragged the piece of iron after him all day. When he was a year and a half old he flew away, with the chain and iron attached to his leg, and perched on the spire of the church of St. Thomas, whence he was scared away by the carrier hawks. On alighting in the street a negro attempted to catch him for the purpose of bringing him home; upon which he seized the poor creature by the ear, and tore it completely off. He then attacked a child in the street, a negro boy of three years old, threw him on the ground, and knocked him on the head so severely, that the child died in consequence of the injuries. I hoped to have brought this bird alive to Europe; but after being at sea two months on our homeward voyage he died on board the ship, in the latitude of Monte Video.'

Humboldt devoted much time and attention to this remarkable bird. It had been represented as far more gigantic than he found it to be in reality; and he himself got an exaggerated impression of its size when he viewed it standing in solitary grandeur on some of those mountain peaks in the regions of perpetual snow.

No bird mounts so high in the air as does the condor. Humboldt says that it will rise 20,000 feet above the sea, and that though the air at that height is so highly rarefied, yet it will descend in a few minutes to the edge of the ocean. Yet the bird enjoys life just where some time back it was supposed that no animal could exist. It only descends to obtain the carrion on which it subsists, and then so gorges itself that it cannot fly. Then it is that the Indians appear with their lassos, and an exciting chase begins which is regarded as only inferior to a bull-fight in interest.

'In riding through the plain,' says Sir Francis Head, 'I passed a dead horse, about which were forty or fifty condors, many of which were gorged and unable to fly; several were standing on the ground devouring the carcase, the rest hovering above it. I rode within twenty yards of them; one of the largest birds was standing with one foot on the ground and the other on the horse's body, with a display of muscular strength as he lifted the flesh and tore off great pieces, sometimes shaking his head, and pulling with his beak, and sometimes pushing with his leg. Got to Mendoza and went to bed. Wakened by one of my party who arrived; he told me that seeing the condors hovering in the air, and knowing that several of them would be gorged, he had also ridden up to the dead horse, and that as one of these enormous birds flew about fifty yards off, and was unable to go any further, he rode up to him, and then jumping off his horse seized him by the neck. The contest was extraordinary, and the *rencontre* unexpected. No two animals can be imagined less likely to meet than a Cornish miner and a condor; and few could have calculated a year ago, when the one was hovering high above the snowy pinnacles of the Cordillera, and the other many fathoms beneath the surface of the ground in Cornwall, that they would ever meet to wrestle and hug upon the wide desert plain of Villa Vicencia. My companion said he had never had such a battle in his life; that he put his knee upon the bird's breast, and tried with all his strength to twist his neck; but that the condor objecting to this struggled violently, and that also, as several others were flying over his head, he suspected they would attack him. He said that at last he succeeded in killing his antagonist, and with great pride he showed me the large feathers from his wing; but when the third horseman came in he told us that he had found the condor in the path, but not quite dead.'

The condor, however, is not the only representative of the vulture tribe in South America. Common vultures abound; and one called the king vulture is also a native, not of the mountains, but of the lower regions. It is the most elegant species in the family, but one of the smallest; and it has therefore obtained its name with no reference to its size, but from the fact that other vultures evidently treat it with respect, and yield it precedence on all occasions. This fact has been noticed everywhere, and thus in Paraguay and in Guiana the names given by the natives signify the same thing.

The bird measures about two feet and a half in length, and double that to the extent of its wings. The colouring of the king vulture is much gayer than that of the condor. Orange and violet tinge the head; there is a loose comb of bright orange over its beak. The iris is of a pure white, with a scarlet circle round the eye, and the sides of the head are purplish black, with a patch of blackish down on the back. Then there is a reddish-brown fold, with some mixture of blue, passing down the neck. Bright red, orange and blue, may be noticed in the neck. The quill-feathers of the wings and tail are black, and so are the feet.

If birds have any eye for beauty, we may understand why this is acknowledged as king of the tribe; and that he is so regarded is evident from the following story related by Waterton. He says that, in order to tempt the royal bird, he had the body of a large serpent which he had killed carried into the forest, and then watched for the result, which he thus describes:—'The foliage of the trees where we laid it was impervious to the sun's rays, and, had any vultures passed over that part of the forest, I think I may say with safety that they could not have seen the remains of the serpent through the shade. For the first two days not a vulture made its appearance at the spot; though I could see here and there, as usual, a *vultur aura* gliding on apparently immovable pinions at a moderate height over the tops of the forest-trees; but during the afternoon of the third day, when the carcase of the serpent had got into a state of putrefaction, more than twenty of the common vultures came and perched upon the neighbouring trees; and the next morning, a little after six o'clock, I saw a magnificent king of the vultures. There was a stupendous mora tree close by, whose topmost branch had either been killed by time or blasted by the thunder-storm. Upon this branch I killed the king of the vultures, before it had descended to partake of the savoury food which had attracted it to the place. Soon after this, another king of the vultures came, and after he had stuffed himself almost to suffocation the rest pounced down upon the remains of the serpent, and stayed till they had devoured the last morsel.'

The same traveller says, that while travelling up the Essequibo he observed a pair of these birds sitting on the naked branch of a tree, with about a dozen of the common species, waiting to begin the feast on a goat which a jaguar had killed the day before; and still, though tolerating the company of the inferiors, appearing to guard their royal privileges with jealous care.

Some eagles also are found in tropical America. There is one kind known in Brazil as the Caracara eagle, from its peculiar cry. It ranges over most parts of South America, is a very fine bird, but in some respects—that is, in the bare appearance of the head, its prominent crop, and the position of the eye—this species resembles the vulture. The colouring of the bird is brownish-grey, brown, and black; there is a little red in the head, and the legs are yellow. It builds on the tops of trees, preferring those which are covered with creepers, or in some bushy thicket. The female is larger than the male; and she lays two eggs of a dull red, dotted with crimson.

Eagles belong to the class which is possessed of the organs of destruction in the highest degree, and which therefore preys chiefly, not on dead, but on living animals. They attack their victims in the air, or on the ground; preferring the latter, as they have not the power of very swift flight. 'The feet are eminently formed for striking and trussing, and the beak for dissecting their prey, which, with hardly any exception, consists of living animals; and for the pursuit and conquest of these the birds are endowed with vigorous limbs, the wings being for the most part long, dense, and capable of powerful flight, and the feet strong and muscular, armed with powerful talons.' This is Mr. Gosse's description, and he adds that in a natural state these birds will never touch any but carnivorous food, but that it has been proved that there is nothing to prevent the digestion of other and vegetable food.

The head feathered, except the cere at the base of the beak; the strong, hooked beak, with a sort of tooth on either side; the nostrils more or less rounded, and pierced in the sides of the cere; the stern, projecting eyebrows; and the strong, sharp talons; these, he says, are the marks of the family. And eagles possess all these marks in the highest degree: but the one of which we have been speaking, the caracara, has just those differences which have been mentioned, and which seem to connect it with the vulture family. It eats, however, both the dead and the living; and sometimes, when one bird cannot take some destined victim, four or five will join to attack it.

A huntsman is sometimes balked of his game because the caracara is beforehand with him, and takes it before his face. And it will sometimes happen that this eagle will pursue a vulture and force him to disgorge his prey.

Mr. Edwards, who in 1846 made a voyage up the Amazon, also speaks of these caracara eagles as common all over the country, and as allied to the vulture. But he also met with another kind of eagle, the harpy.

A number of singular birds and curiosities, he says, were brought by a trader to the house where he was staying. 'One of the former was a young harpy eagle, a most ferocious-looking character, with a harpy's crest and a beak and talons in correspondence. He was turned loose into the garden, and before long gave us a sample of his powers. With erected crest and flashing eyes, uttering a frightful shriek, he pounced upon a young ibis, and, quicker than thought, had torn his reeking liver from his body. The whole animal world below there was wild with fear. The monkeys scudded to a hiding-place, and parrots, herons, ibises, and mutins, with all the hen tribe that could muster the requisite feathers, sprang helter-skelter over the fences, some of them never to be reclaimed.'

Another time he says, 'Just before we reached our mooring a full-sized harpy eagle perched upon a tree near the water, his crest erect, and his appearance noble beyond description. We gave him a charge of our largest shot, but he seemed not to notice it. Before we could fire again he slowly gathered himself up and flew majestically off. This bird is called the Gavion Real, or Royal Eagle, and is not uncommon throughout the interior. Its favourite food is said to be sloths and other large-sized animals.'

Leaving now for awhile the mountain-peaks, the pampas and llanos, and other inland situations, let us inquire a little into the coast-birds of this part of our globe. 'Three of the most remarkable of these wild wanderers,' says Stanley, 'are the albatross, the tropic bird, and the frigate-bird. The first of these, the albatross—the largest of the aquatic tribe—with plumage of the most delicate white, except the back and tops of its wings, which are of a dark grey, floats in the air borne up by a vast expanse of wing, measuring fourteen feet, or even more, from tip to tip. The air and the water, indeed, seem to be far more natural to it than the land, where it is so helpless, owing to its enormous length of wing, which prevents it from rising unless it can launch itself from a steep precipice or projecting rock, that it is completely at the mercy of those who approach, and one blow on the head generally kills it instantly.' The size of the bird itself is not, however, so large as might be supposed, so much of its magnitude being due to the quantity and size of its feathers. They are, indeed, very light, and this circumstance, added to their immense extent of wing, accounts in some measure for their being able to venture so far out to sea.

It is a well-established fact that these magnificent birds will frequently follow the course of ships for many miles; and very welcome must their company often be in the midst of the vast solitude. They inhabit the entire circle of the globe, keeping to the southern seas, and are found sometimes, in company with the penguins, on some barren rock or desert island.

'It has been remarked that these birds could lower themselves, even to the water's edge, and then again rise without any apparent impulse; whether with or against the wind, seems to be a matter of indifference to them. No tempest troubles the albatross, for he may be seen, with equal vigour, sportively wheeling in the blast and carousing in the hurricane. Of this noble bird it may indeed be literally said,—

"His march is o'er the mountain wave,
His home is on the deep."

In the gale he will sweep occasionally the rising billows, and seem to delight in the spray bursting over him. Tired, in truth, they rarely are; but should they be, though never seen to swim, they can, in consequence of their feet being webbed and remarkably large, walk on the surface of the water when it is smooth, with hardly any assistance from their wings; and the noise of their tread may be heard at a great distance. They are most voracious birds, and easily caught by baiting a hook with offal and letting it trail after the vessel by a long line; on seizing and swallowing the bait it will sometimes rise into the air, from whence, by hauling in the line, as a boy does a kite, it is brought on board. Sometimes, however, they will break the line and escape. . . . Thus, when hauling one in of large size, the line slipped, and the bird consequently swallowed the hook, and a portion of the line, the remainder of which hung pendant from the beak. . . . Their reason for preferring rough weather to smooth may easily be accounted for, the agitation of the waves no doubt bringing to the surface those marine

THE ALBATROSS.

animals which serve them for food: they will glide down on them with unerring aim and fearful force, transfixing whatever they have aimed at with their large, strong, and trenchant bill. A poor fellow, who fell overboard from a man-of-war off the Island of St. Paul's, in the Southern Indian Ocean, was immediately perceived by two or three albatrosses; the boat was lowered with all speed, but nothing was found excepting his hat, pierced through and through with the violent stroke

MARINE BIRDS.

of their beaks, the first of which had, most probably, penetrated the skull, and caused instant death.'

Albatrosses are very fond of blubber, and often gorge themselves till they cannot fly. A flock of five or six hundred of them have been known to eat up twenty tons of sea-elephant fat in six or eight hours; that is, seventy pounds for each. One of them will swallow at one gulp a whole salmon of about five pounds' weight; and if he takes too much, some of the fish remains hanging out of his mouth.

Sailors accustomed to these latitudes are familiar also with another bird of kindred habits. This is the Tropic bird, known by the length of its wings and its two tail feathers, so long and slender—a bird which but seldom comes to land, except at night and to lay its eggs, when it perches for the time on some rock or tree, and is off again as soon as possible. The albatross is heavy and massive, the tropic bird delicate and fairy-like; and often may it be seen as if it were a tiny speck at rest in the

FLYING-FISH.

sky, then darting like a meteor on to its prey, and rising again with its prize up amongst the clouds. This bird is about the size of a partridge.

Then we have the Frigate bird, one of the most singular of the feathered race, with its partially webbed feet, and its half-inch legs covered with the long loose feathers, its long forked tail, and its marvellous spread of wings, feathered, but not with the close, downy texture which other birds have that skim the deep continually. No bird can equal this in powers of flight. The frigate birds go in flocks, and have a noble appearance as they perform their dexterous evolutions, and dart skilfully on their prey; they do this as soon as the substance is visible, for they cannot see many inches below the surface. Mr. Darwin says that he saw them frequently descend for their prey, and then rise quietly and gracefully, and that they seldom missed their aim. 'I was informed at one place,' he says, 'that when the little turtles break their shells and run to the water's edge, these birds attend in numbers and pick up the little animals from the sand, in the same manner as they do from the sea.'

The frigate bird, like the albatross, is seen far out at sea, but ever on the wing, for its structure unfits it for a life either in the water or on the land; indeed its wings are so long that, except from some rock or projecting point, from which it can float into the air, it can scarcely take flight. From tip to tip they measure twelve feet, and by constant habit it can keep them always extended. Then, just beneath the throat, there is a large pouch communicating with the lungs, and with the hollow and very light bone-work of its skeleton. This is an apparatus for keeping the bird floating in the atmosphere. He can at will fill this pouch with air, which by the heat of his circulation becomes rarefied, and fills all the cavities. He cares nothing, therefore, for storms or fogs, for whenever these come on up he floats to a clearer space. Then, when hunger compels him to descend, he empties this pouch, and down he dives after the flying-fish, which he catches with his bill or talons.

Herons are found all over the world, though but seldom in the colder parts. They are the most beautiful of the Waders. They watch patiently by the water-side for their prey, and as soon as it appears instantly transfix it with a lightning stroke, and swallow it whole.

Some very large herons are found in South America. Dr. Masterman speaks of finding a fine specimen of the great heron, nearly as tall as himself, with a bill nearly a foot in length. He kept him tied with a rope, to which a heavy paving-brick was fastened. One day he was frightened, and flew off, brick and all; the brick striking the wall, and thus breaking off, nearly killed a soldier who was lying asleep under it. The great white heron is found in Guiana. These birds build in companies, and on the trees in low grounds. Their plumes have at certain times been much in request in Europe for head-dresses.

Edwards says that, during his passage up the giant river of South America, 'sometimes, far in the distance, the keen eyes of the men would descry the great blue heron, the Ardea Herodias, and with silent oars and beating hearts we crept along the shore, hoping to take him unawares. But it was of no avail; his quick ear detected the approaching danger, and long before we could attain shooting distance he had slowly raised himself and flown further on, only to excite us still more in his pursuit.'

Tropical travellers frequently meet with the Toucan : a bird remarkable for his very long beak, the short, rounded wings, long and broad tail, and feet formed for grasping. Mr. Waterton speaks of three varieties, besides smaller birds, which he calls toucanets. The largest feeds on the mangrove-trees on the sea-coast; the other two kinds feed entirely on the fruits of the forests, and, though of the pie kind, never kill the young of other birds or touch carrion. They like feeding in company on the same tree, and then afterwards go away together to some shady spot in the forest. The Indians call the larger kind bouradi, but the Spaniards piapoco, from his peculiar cry, which is like the yelping of a puppy-dog. All the toucanets feed on the same trees on which the toucan feeds, and every species of this family of enormous bill lays its eggs in the hollow trees. They are social, but not gregarious. You may sometimes see eight or ten in company, but you will find it has only been a dinner-party, which breaks up and disperses towards roosting-time. You will be at a loss to conjecture for what ends Nature has overloaded the head of this bird with such an enormous bill. It cannot be for the offensive, as it has no need to wage war with any of the rest of animated nature, for its food is fruit and seeds, and there are a superabundance throughout the whole year in the regions where the toucan is found.' Stanley thus speaks of the toucan's enormous beak :—' In the toucan the beak forms a most prominent and unsightly feature, being quite a deformity in the otherwise beautiful and graceful bird; and were it as heavy in proportion as the bills of other birds, it might prove a very serious weight, and materially impede his flight, if not quite weigh it down to the ground. It is, however, so remarkably light and hollow as to be no inconvenience whatever, so that the bird can fly with such swiftness and certainty as to catch grapes and other fruits thrown to it before they fall to the ground. In its operation, too, it differs from those of other birds : seizing and acting upon the substances within its grasp by a lateral or side-way, rather than up-and-down or perpendicular motion. But they do not always confine themselves to fruits, their beaks being equally calculated by their muscular strength for crushing the bones of small birds ; and in their native forests they are seen perched on high trees watching the moment when old birds leave their nests, when down they pounce and feed on the young ones, and even contest a prize with the monkeys. How skilfully, and at the same time how powerfully, he can use this apparently awkward and cumbrous bill of his, we learn from the way in which a toucan, which was kept for some years in the Museum of the Zoological Gardens in London, disposed of a small bird. The moment the owner of the toucan introduced his hand with the small bird into the cage, the toucan, which was on its perch, snatched it with its bill. The poor little bird was dead in an instant, killed by the violence of the squeeze. As soon as it was dead the toucan hopped, with it still in its bill, to another perch, and then, placing it with his bill between his right foot and the perch, began to strip off the feathers. When he had plucked away most of them he broke the bones of the wings and legs (still holding the little bird in the same position) with his bill, taking the limbs therein, and giving at the same time a strong lateral wrench. He continued this work with great dexterity till he had almost reduced the body to a shapeless mass. He at first ate all the soft parts, leaving the larger bones to the last, which seemed to give him more trouble, particularly the beak and legs.'

Other naturalists have doubted the carnivorous nature of these birds, but certainly this story seems to establish it as a fact. Mr. Jesse believes that the toucan defends itself against the monkey tribe by means of its bill. He says, too, 'the toucan feeds on the eggs of other birds. By means of its long bill it is enabled to search for and reach them in the holes of trees, and also in the pendent nests of tropical birds. These nests are suspended from the extremity of branches of trees, as a security against monkeys, and have a hole in the side, into which the toucan is enabled to thrust its long bill.'

There is a small variety of the toucan which has on its head 'thin horny plates, of a lustrous black colour, curled up at the ends and resembling shavings of steel or ebony, the curly crest being arrayed on the crown in the form of a wig.' Mr. Bates having wounded one of these birds, on his attempting to seize it, it set up a loud scream. 'In an instant, as if by magic, the shady nook seemed alive with these birds, though there were certainly none visible when I entered the jungle. They descended towards me, hopping from bough to bough, some of them swinging on the loops and cables of woody lianas, and all croaking and fluttering their wings like so many furies.'

When settling itself to sleep the toucan packs itself up, supporting its huge beak on its back, and tucking it amongst the feathers, while it doubles its tail across its back, so that altogether the bird looks like a great ball of loose feathers.

'Among the natural productions which I saw here for the first time,' says Prince Maximilian, 'the emu, or nandu, is not the least interesting. This great bird, which is the ostrich of America, is very common in the Campos Geraës, where it is seldom hunted. A female with its little ones, which had been hatched about six months, was living quietly in the neighbourhood of Nalo. No one disturbed it until some greedy Europeans came to trouble its peace and attempt its life. This bird, very suspicious and cunning, perceives the hunters at a great distance; much precaution is therefore necessary in order to take it. In running it tires out a horse, because it does not go in a straight line, but makes many turns. When the nandu, with her fourteen little ones, which were half their full size, showed herself for the first time, after we had watched for her in vain for several days, three of my hunters placed themselves in ambuscade, and we attempted to drive the nandus towards them; but the birds were as cunning as we, and did not allow themselves to be deceived. At this moment a vaquero on horseback, and well armed, chanced to arrive, and at once resolved to catch the birds. He began by following the pack slowly, then at full gallop, and after many attempts succeeded in taking one of the little ones by quickly jumping off his horse. A well-aimed fire with large shot brought down the largest. We frequently followed this kind of chase, and one of my hunters, towards whom we had driven three nandus, killed an old one: it was a hen. She was four feet, four inches from the top of the beak to the end of the tail, and she weighed $56\frac{1}{2}$ lbs. I found in her stomach small cocoa and other hard fruits, and all kinds of grass, the remains of snakes, grasshoppers, and other insects. The flesh of the nandu has a disagreeable smell, and is not eaten. It is said to follow dogs. In these parts its skin is tanned and dyed black, and employed for making garters, on which the marks of the feathers may be seen. Purses are made of the long skin of the neck; the eggs divided, in the middle, serve for *coins au jattes*, and the feathers for fans.'

Stanley says that 'the American ostriches are not only most affectionate, but sociable, laying together in the same nest, or rather the same hole, showing equal attention to their joint broods.' He also tells an affecting tale. 'A pair of ostriches had long been kept in the Zoological Gardens at Paris. The skylight over their heads having been broken, the glaziers proceeded to repair it, and in the course of

HUNTING EMUS ON THE CAMPO.

their work let fall a triangular piece of glass. Not long after this the female ostrich was taken ill, and died in an hour or two in great agony. The body was opened, and the throat and stomach were found to have been dreadfully lacerated by the sharp corners of the glass which she had swallowed. From the moment his companion was taken from him the male had no rest; he appeared to be incessantly searching for something, and gradually wasted away. He was moved from the spot, in the hope that he would forget his grief; he was even allowed more liberty: but nought availed, and he literally pined away till he died.'

The Cassowary is of about the same size as the emu, and like it resembles the ostrich in general appearance and habits. Its neck and legs are shorter than the ostrich, and its form is more clumsy. Dogs are very shy of hunting it on account of the violent kicks which it can give with its stout legs. It will outstrip the fleetest horse in running, and on occasion will take to the water.

In South America there are several birds whose notes are so singular that the natives look on them with a kind of reverence, and will never kill them. One of the most remarkable is that called the Campanero by the Spaniards, Araponga by the Indians, and Bell-bird by the English. It is about the size of a jay, and the plumage is a pure white. 'On the forehead,' says Waterton, 'rises a spiral tube, nearly three inches long. It is jet-black, dotted all over with small white feathers. It has a communication with the palate, and when filled with air looks like a spire; when empty it becomes pendulous. His note is loud and clear, like the sound of a bell, and may be heard at the distance of three miles. In the midst of these extensive wilds, generally on the dead top of an aged mora, almost out of your reach, you will see the campanero. No sound or song from any of the winged inhabitants of the forest, not even the clearly-pronounced 'Whip-poor-Will,' from the goat-sucker, cause such astonishment as the toll of the campanero.

'With many of the feathered race he pays the common tribute of a morning and evening song, and even when the meridian sun has shut in silence the mouths of almost the whole of animated nature, the campanero still cheers the forest. You hear his toll, and then a pause for a minute, then another toll, then a toll, and again a pause. Then he is silent for five or six minutes, then another toll, and so on.

'Actæon would stop in mid chase, Maria would defer her evening song, and Orpheus himself would drop his lute, to listen to him; so sweet, so novel and romantic, is the toll of the pretty snow-white campanero!'

This bird is found also in Africa, but its note has not the same cheering effect on every one. Two missionaries were once journeying onwards in the solitude of the wilderness, when the note of the campanero fell upon their ear. '"Listen," said the companion of the one who relates the circumstance; "did not you hear a church-bell?" We paused, he says, and it tolled again, and so strong was the resemblance, that we could scarcely persuade ourselves that we did not hear the low and solemn sound of a distant passing-bell. When all was silent it came at intervals upon the ear, heavy and slow like a death-toll. All again was then silent, and then again the bell-bird's note was borne upon the wind. We never seemed to approach it, but that deep, melancholy, distant, dream-like sound, still continued at times to haunt us like an omen of evil.'

Edwards, again, describes it thus: 'During the night we fancied we heard the far-famed bell-bird. The note was that of a muffled tea-bell, and several of these ringers were performing at the same time; some with one gentle tinkle, others with a ring of several notes. I asked the pilot what was "*Gritando?*" He replied, "A toad." I had no idea of having my musician thus calumniated, and remonstrated thereupon; but he cut me short with "It must be a toad; everything is a toad that sings at night!"'

Another bird, nearly allied to the campanero, is the Tanager, celebrated for its very brilliant plumage. In the different species every colour in its brightest hue may be found, and sometimes mingled together as in the painted tanager, where the brightest shades of green, blue, orange, and black, are so intermingled as to make the feathered creature quite dazzling. Edwards met with the scarlet variety, 'whose black-masked, brilliant, metallic, scarlet and black, livery, was like a jewel in the sunlight. We had seen nothing comparable to it on the river,' he adds. 'These birds were always seen about low bushes by the water-side, catching their favourite insects, and uttering a slight note or whistle.'

The Cassique is another remarkable bird. Waterton says, 'It is larger than a starling, and that it courts the society of man, but disdains to live by his labours. When nature calls for support he repairs to the neighbouring forest, and there partakes of the store of fruits and seeds which she has produced in abundance for her aerial tribes. When his repast is over he returns to man, and pays the little tribute which he owes him for his protection. He takes his station on a tree close to his house; and there, for hours together, pours forth a succession of imitative notes. His own song is sweet, but very short. If a toucan be yelping in the neighbourhood, he drops it and imitates him. Then he will amuse his protector with the cries of the different species of woodpeckers, and when the sheep bleat he will distinctly answer them. Then comes his own song, and if a puppy-dog or a guinea-fowl interrupt him, he takes them off admirably, and by his different gestures during the time you would conclude that he enjoys the sport.

'The cassique is gregarious. He goes by the name of the Mockingbird among the colonists. At breeding-time a number of these pretty choristers resort to a tree near the planter's house, and from its outside branches weave their pendulous nests. So conscious do they seem that they never give offence, and so little suspicious are they of receiving any injury from man, that they will choose a tree within forty yards of his house, and occupy the branches so low down that he may peep into the nest. The proportions of the cassique are so fine, that he may be said to be a model of symmetry in ornithology. On each wing he has a bright yellow spot, and his rump, belly, and half the tail, are of the same colour. All the rest of the body is black.'

Another strange native of Brazil is the Cephalopetrus ornatus, or Umbrella bird. It is one of the most curious and most rarely met with. 'We saw a pair of these umbrella chatterers,' says one of the travellers so often quoted; 'they were sitting near together upon the lower branches of a large tree, and a shot brought down the female. Unfortunately the gun had been loaded but in one barrel, and before ammunition could be obtained from the boat the male, who lingered about for some moments, had disappeared. We afterwards obtained a fine male upon the Rio Negro. These birds are of the size of small crows, and the colour of their plumage is a glossy blue-black. Upon the head is a tall crest of slender feathers, whence it derives its name, and upon the breast of both male and female is a pendant of feathers of three inches. They are, like all the chatterers, fruit-eaters. They are pretty common upon an island a few days' sail above the barra of the Rio Negro, but they are not found anywhere in that region in such flocks as others of the chatterer family. The Indian name for these birds is *uira-mimbeu*, and the taucha informed us that they built in trees and laid white eggs.'

In building their nests all these birds, and most others, have an eye to the various enemies by whom they are surrounded; as, for instance, lizards and other reptiles, and above all of the white ant, which infests every spot. And many a nest is arched over to keep it from the excessive heat of the sun. The most singular are the nests of the Troupials, a large black bird much marked with yellow. Their native name is Japim. They build in colonies pensile nests of grass, nearly two feet in length, having an opening for entrance near the top.

It is superfluous to say that the large order of Climbing-birds, to which parrots belong, is well represented in South America. The distinguishing marks of this order are found in their beaks and feet. The beaks are large, with the upper mandible much curved over the lower, and very powerful. The tongue is short, thick, and fleshy, suited for articulation. For the feet, two claws run forwards and two backwards, enabling the bird to cling and climb well. The true parrot's tail is square and short, but those of paroquets and macaws are very long.

Cockatoos are Australian birds; but macaws are peculiarly American. They are amongst the most gorgeous of the whole family, and large in size. Quite in the south there is a bird of this family called the Arara. Waterton speaks of the macaw as the ara, and says, referring to the green, blue, and yellow macaw, 'It will force you to take your eyes off any other bird to look at him. This kind fly high, and may be seen by thousands overhead, with their scarlet bodies, and their red, blue, and yellow wings. They go in pairs, and are easily tamed. They can learn to talk a little, but are among the most noisy of the parrot-kind. The great green macaw differs from the blue and yellow in its habits, the former loving the extensive maize-fields, while the former enlivens the forests. There is also the great scarlet macaw, and another species which is red and blue, as well as one called the noble parrot macaw.'

In his account of his voyage up the Amazon Edwards writes: 'The woods about Taüaü were of the loftiest growth, and filled with game, both birds and animals. Here we first encountered the gorgeous macaws, climbing over the fruit-covered branches and hoarsely crying. They were wiser than most birds, however, having gained caution by long experience; for their brilliant colours and long plumes render them desirable in the eyes of every Indian. They were not unwilling to allow us one glimpse, but beyond that we never attained.' But again, he says, 'Towards evening we came to a place where the macaws were assembling to roost. Disturbed by our approach they circled over our heads in great numbers, screaming outrageously. A—— caught a gun, and as one of them came plump into the water, men, women, and children set up a shout of admiration. Two of the boys were instantly in the stream in chase of the bird, which was making rapid strokes towards a clump of bushes. Macaw arrived first, and for joy at his deliverance, laughed in exultation; but a blow of a pole knocked him into the water again and a towel over his nose soon made him prisoner upon our own terms. The poor fellow struggled lustily, roaring and using bill and toes to good purpose. His sympathizing brethren flew round and round, screaming in concert; and it was not until another shot had cut off the tail of one of the most noisy that they began to credit us for being in earnest. Our specimen was of the blue-and-yellow variety. During the night we repeatedly sailed by trees where those birds were roosting, and upon one dry branch A——, whose watch it was, counted eighteen.' Further on he saw a pair of hyacinthine macaws, entirely blue, the rarest variety upon the river. In some parts, too, there are paroquets and parrots, all of brilliant hues.

In America alone and its islands have Humming-birds been as yet found. They are the tiniest and yet most brilliant of all the feathered tribes, so metallic and burnished are their tints. The sun-birds of the old world strongly resemble them, though they scarcely equal them in loveliness, and, moreover, are larger birds. Some humming-birds are scarcely larger than a good-sized hornet; and three inches in length is an average size, though it is true that a few are considerably larger. The name is derived from the noise they make with their excessively long wings; but the noise varies in each species, and there are 300 sorts. Their legs are so weak that they perch comparatively seldom, while their wings are very strong and their flight most rapid.

Both sun-birds and humming-birds are honey-suckers; but the latter dazzling little winged mites often suck the nectar from the blossom without touching it except with their beaks. The senses of touch and smell are evidently very highly developed in them, for from the length and form of their beaks it is impossible for them to see the flowers whence they are drawing the nectar. A long double tongue, which they can push out a long way, effectually does their work.

But the humming-birds love small insects also, and are expert fly-catchers. They are chiefly found in the tropics, but are not exclusively lovers of the sunshine, as many flit about in the eternal twilight of their vast native forests.

These exquisite little feathered fairies are as swift of wing as they are dazzling in tint, whizzing past you one moment and out of sight the next. No bird of prey can come near them; they can distance the swiftest in an instant. They fear neither the fiercest nor the most powerful; on the contrary, when offended, one of these little creatures will attack the proudest eagle, alighting on his head in its rage, and pecking away at his feathers with all its might, the royal bird can do nothing to get rid of his tormentor, but has to bear his persecution just as long as he chooses to stay.

Mr. Webber, the author of a book on song-birds, succeeded in securing an uninjured captive, which, 'to my great delight,' he writes, 'proved to be one of the ruby-throated species. It immediately suggested itself to me that a mixture of two parts refined loaf-sugar with one of fine honey and ten of water would make about the nearest approach to the nectar of flowers. While my sister ran to prepare it, I gradually opened my hand to look at my little prisoner, and saw, to my no little amusement, as well as suspicion, that it was actually "playing possum," feigning to be dead most skilfully. It lay on my open palm motionless for some minutes, during which I watched it in breathless curiosity. I saw it gradually open its bright little eyes to peep whether the way was clear, and then close them slowly as it caught my eyes upon it. But when the manufactured nectar came, and a drop was touched upon the point of its bill, it came to life very suddenly, and in a moment was on its legs, drinking with eager gusto of the refreshing draught from a silver tea-spoon. When sated, it refused to take any more, and sat perched, with the coolest self-composure, on my finger, and plumed itself quite as artistically as if on its favourite spray. I was enchanted with the bold, innocent confidence with which it turned up its keen black eyes to survey us, as much as to say, "Well, good folks, who are you?" The little bird, however, was quite tamed by the next day, and would perch on the tea-cup and chat away in its own style, as if aware that it was with friends.' Mr. Webber thus succeeded in taming several, and the next spring they came back to the very same window, bringing their mates and friends with them, and flying fearlessly in at the casement to seek the well-remembered feast.

These Ruby-throats have a curious way of flying straight up in a perpendicular line, away from their nests, and in returning they appear to drop down into them in the same way. They thus keep the position of their nests a secret, and beautifully delicate constructions these nests are. They make them of the lichen that grows on the same bough to which they attach them, first lining them with cottony substances and then with a silky coating. In the nest so prepared the hen lays two white oval eggs. They are hatched in ten days, and in a week ready to fly, but the parents feed the young birds for two weeks; after which time they leave them to provide for themselves.

'Never,' says Captain Head, 'was I more excited to wonder than by one of these little creatures, so much more resembling a splendid shining insect than a bird. It was on a fine day at the commencement of an American summer, on the banks of Lake Huron, that I first beheld them. Beautiful birds were drinking and splashing themselves in the water, and gaudy butterflies, of a very large size, were fanning the air with their yellow and black wings. At this moment a little blazing meteor shot, like a glowing coal of fire, across the glen, and I saw for the first time, with admiration and astonishment, what in a moment I recognised, that resplendent living gem, the humming-bird, buzzing like a humble-bee, which it exactly resembled in its flight and sound. Like it, it sprang through the air by a series of simultaneous impulses, tracing angle after angle with the velocity of lightning, till, poised above its favourite flower, all motion seemed lost in its very intensity, and the humming sound certified to the ear the rapid vibration of its wing, by which it supported its little airy form.'

They vary from the size of a humble-bee to that of a willow-wren; the nests of the smaller sort not exceeding an inch in diameter, and formed of the most delicate materials, appearing more like mossy knots on a branch than the manufacture of a bird. They will build fearlessly within sight of a window, where they may be leisurely observed. They frequently assemble in great numbers round some sorts of flowers yielding those sweet juices, which, together with insects, compose their food. The aloe is one of them.

A gentleman in Jamaica thus describes them hovering round a plot of these plants, covering nearly twenty square yards, of which about a dozen were in full bloom:—'The spikes bearing bunches of flowers were from twelve to fifteen feet high; on each spike were many hundred blossoms of a bright yellow colour, each of a tubular shape, and containing its drop of honey. These alone afforded,' as he says, 'a splendid scene, but the interest was doubled by the addition of vast numbers of humming-birds fluttering round the openings of the flowers, and dipping their bills first into one flower and then into another, the sun shining bright upon their beautiful plumage, giving them the appearance of now a ruby, then a topaz, then an emerald, and then all burnished gold.'

But there are other sorts of oviparous animals in these countries, and far less attractive ones, than those of which we have been speaking.

Some of the rivers abound with alligators, which in the daytime lie basking in the sun with their mouths open. Into these wide cavernous jaws there fly numerous insects, allured by the odour of their breath, and from time to time they swallow the whole multitude. The female lays her eggs in the sand, depositing from 80 to 100 in one or two days. These eggs are longer than the eggs of geese, and although tough, yet they are tolerable eating. When instinct informs the alligator that the time of ovation is completed, both the male and the female go to the nest, and, if undisturbed, the female immediately uncovers the eggs and carefully breaks them; the young brood begin to run about, and the watchful gallinasos prey upon them; while the male alligator, who appears to have come for no other purpose, devours all that he possibly can: those that can mount on the neck and back of the female are safe, unless they happen to fall off, or cannot swim, in which cases she devours them. Thus Nature has prepared a destructive for these dangerous animals, which would otherwise be as numerous as flies, and become the absolute proprietors of the surrounding country: even at present, notwithstanding the comparatively few that escape, their number is almost incredible. They feed chiefly on fish and young animals, but when once they have tasted human flesh they become dangerous. Indeed, when they find calves and fowls easy to procure they then give up the fish with which they had been content. The natives consider it good sport to catch these creatures by means of a baited noose, which allures the animal within reach of their lances.

At other times they take them alive, and one method of doing so is as follows:—' A man takes in his right hand a truncheon, called a *tolete*; this is of hard wood, about two feet long, having a ball formed at each end, into which are fastened two iron harpoons, and to the middle of the truncheon a platted thong is fastened. The man takes this in his hand, plunges into the river, and holds it horizontally on the surface of the water, grasping a dead fowl with the same hand, and swimming with the other; he places himself in a right line with the lagarto, which is almost sure to dart at the fowl: when this happens, the truncheon is placed in a vertical position, and at the moment that the jaw of the lagarto is thrown up the tolete is thrust into the mouth, so that when the jaw falls down again the two harpoons become fixed, and the animal is dragged to the shore by the cord fastened to the tolete.

When on shore the appearance of the lagarto is really most horrible; his enormous jaw, propped up by the tolete, showing his large sharp teeth; his eyes projecting almost out of his head; the pale red colour of the fleshy substance on his under-jaw, as well as that of the roof of the mouth; the impenetrable armour of scales which covers the body, with the huge paws and tail, all contribute to render the spectacle appalling: and although one is perfectly aware that in its present state it is harmless, yet it is almost impossible to look on it without feeling what fear is. The natives now surround the lagarto and bait it like a bull, holding before it anything that is red; at which it runs, when the man jumps aside and avoids being struck by it, while the animal continues to run forwards in a straight line till checked by the thong which is fastened to the tolete. When tired of teasing the poor brute, they kill it by thrusting a lance down its throat, or under the fore-leg into its body, unless by accident it be thrown on its back, when it may be pierced in any part of the belly, which is soft and easily penetrated.'

Some naturalists consider these creatures the American crocodile; but Humboldt maintains that he saw true crocodiles there in considerable numbers; and he speaks of them as most dangerous, saying that they frequently seize and drag human beings under water, and after having drowned, will devour them at their leisure. He thinks that few people know how many lives are thus constantly lost.

Speaking of Bahia Blanca, Edwards says, that 'of reptiles there are many kinds;' and he describes one which appears to be a kind of rattlesnake; and states that as often as the animal was irritated or surprised its tail was shaken, and the vibrations were extremely rapid; in some respects its structure was that of the viper, and the noise which it made was produced by a simpler device than that of the rattlesnake. The expression of this snake's face was hideous and fierce. 'I do not think that,' he adds, 'I ever saw anything more ugly, except perhaps some of the vampire bats.'

The true Rattlesnake is said not to attack any animals except those on which it feeds; and its bite, though dangerous at certain seasons, is at other times harmless. The rattling noise is produced by a number of articulated cells at the end of its tail, which always give warning of its vicinity.

The Anaconda is a very large snake, found along the Amazon. One which had long infested an estate, and had at different times carried off forty pigs, was captured by means of a lasso. The ant-bear frequently falls a victim to this snake ; yet he is kept in some houses in order to destroy the rats, and is reckoned harmless.

The Boa-constrictor, of which there are four kinds, is a native of inter-tropical America. Those gigantic creatures, which can climb, swim, and dart along the ground, are the terror of man and beast. Its hissing creates an agony of fear as soon as it is heard. This snake frequents marshy places, and kills its victims by folding itself round and crushing them. One of them, which was in the London Zoological Gardens, swallowed a railway wrapper on one occasion instead of the rabbits which had been given him for his supper. He afterwards disgorged it, and fasted for a week ; but eventually he was nothing the worse.

Turtles are especially African, but some kinds are found in South America ; and aquatic tortoises—*i.e.* turtles—are innumerable in the rivers of that country. Edwards speaks of the turtle of the Amazon as most numerous, and excellent eating. He bought one weighing a hundred and twenty-five pounds, and put him in the hold of the vessel. This animal knew how to defend his own dignity, and resented a touch on the head by seizing the mate's thumb and maiming him.

He speaks of these turtles as a great source of support and riches to the natives, and says that in the dry season they come up from the sea and deposit their eggs in the sand. These are picked up and broken into six-gallon pots, and the oil skimmed off. It is computed that a turtle lays a hundred and fifty eggs in a season. Twelve thousand make up one pot of oil ; and six thousand pots are sent from the most noted places every year. Thus seventy-two millions of eggs are annually destroyed, yet in fifty days millions of young turtles are marching to the water, where various creatures impatiently await their coming. Humboldt likewise quotes a statement made by an inhabitant, 'that the shores of the Orinoco contain fewer grains of sand than the river contains turtles, and that these animals would prevent the advance of vessels if men and wild beasts did not annually destroy them.'

'Of lizards,' says a famous traveller, 'there are many kinds, but only one remarkable in its habits. It lives on the bare sand near the sea-coast, and from its mottled colour, the brownish scales being speckled with white, yellowish red, and dirty blue, can hardly be distinguished from the surrounding surface. When frightened, it attempts to avoid discovery by feigning death, with outstretched legs, depressed body, and closed eyes ; if further molested, it buries itself with great quickness

BOA CONSTRICTOR.

in the loose sand. This lizard, from its flattened body and legs, cannot run quickly.'

The largest frogs in the world are found in America, and Darwin mentions one which would appear to be one of the smallest, and sits on a blade of grass about an inch above the surface of the water, sending forth a pleasing chirp. At times several are found together, singing in harmony on different notes. This frog could crawl up a pane of glass.

Frogs and toads abound everywhere in the tropics. There is a singular kind of toad in Guinea and Brazil. As fast as the female lays her eggs the male spreads them over her broad, flat back. Then a number of little pustules arise, which seem to absorb them, one in each pit, till the back looks like a piece of honeycomb. Here the eggs are not only hatched, but the tadpoles turn into complete toads.

Allusion has already been made to the myriads of insects. Some of these cause immense discomfort; as, for instance, the white ants, or termites, which swarm in the western as much as in the eastern tropics, and are as tormenting and destructive as in India. But the black ants are common as well; and the two races have a great antipathy one for the other. The way to destroy the termites is, therefore, to make use of this antipathy, and it is done in this way: as soon as they are observed a little sugar is put down, which instantly attracts the black ones, who attack and drive away the white very speedily.

The Ecitons, or foraging ants, march in columns three or four yards in width, and two or three hundred yards long; they live upon insects, cockroaches, spiders, &c., and rob the nests of another species of their larvæ. Another kind, called the Sauba, mines underground to such an extent that the smoke of a fire lit at one of their entrances has been seen issuing from the ground at half a mile distance.

Where other insects are so numerous, spiders, as might be expected, are numerous likewise. Some of them are gregarious, and construct webs of great extent, which they inhabit in common.

Strange to say, even humming-birds, which do not fear man, and scorn the eagle or the vulture's power, are yet afraid of these spiders. Mr. Badcock watched some stealing the small insects out of the webs, and noticed their ingenuity in managing it. 'They would advance,' he says, 'beneath the web, and enter the various labyrinths and cells, taking care to make good their retreat if the spiders sallied forth to repel them. In ascending some of the spiders' fly-traps great skill and care were required; sometimes the bird had scarcely room for his little wings to spread, and the least mismanagement would have ensnared him in the meshes of the web and insured his destruction. It was only the outposts of the comparatively small spiders, of about his own size, that the humming-bird durst attack, as the larger sort rush out in defence of their property, when the robber would shoot off like a sunbeam, and could be only traced, like an electric spark, by the luminous glow of its refulgent colours.'

These large spiders are sometimes of immense size. Mr. Bates saw some Indian children leading one about the house, as they would a dog, by means of a cord tied round its waist. Some species make their nests in the ground, excavating a hole which they cover with a trap-door so artistically contrived, and so well concealed, that a person who sees one and tries to catch it, may think that his eyes have deceived him, as the door is covered on the outside with earth, and when closed so exactly resembles the surrounding soil as to be undistinguishable.

But there are compensations for these disagreeables. In many places flies are absent, while moths are said not to be so numerous as in England. This refers probably to the small and destructive kind.

Of the butterfly species, Darwin says that they bespeak the zone which they inhabit. In those gorgeous Brazilian forests, where monkeys skirmish through the trees and squirrels leap from bough to bough, while the sloth climbs more at leisure, and the tiny deer bound fearlessly about ; where birds of gaudiest plumage flit, sing, and chatter ; parrots, paroquets, magnificent macaws, woodpeckers, wood-pigeons, thrushes, toucans, and humming-birds, and what not ; there also 'large butterflies float past, the bigness of a hand,' some of a rich metallic blue, some making a 'clicking noise like a toothed-wheel passing under a spring-catch ;' and then, as night comes on, the fire-flies in myriads enjoy their airy dance.'

CHAPTER IX.

VOLCANOES, EARTHQUAKES, METEORS, CAVERNS.

ALTHOUGH volcanoes are not peculiar to the tropical regions, yet they are more numerous in those parts of the globe than in the temperate zones; and about 100 of these wonderful displays of the forces included in the constitution of our earth are contained within the boundaries of South America. Of these 38 are situated in Guatemala, 22 in Chili, and the rest are distributed over the chain of the Andes in Quito, Peru, and Bolivia, all of them being within a comparatively short distance of the coast.

Of these, Chimborazo and Cotopaxi are two of the most remarkable. The former is now extinct, and no tradition exists of its being in an active state; but hot springs on the north side, and the substance with which it is covered, consisting of calcined matter resembling white sand, seem to testify to its volcanic character. Humboldt, however, seems to doubt whether it has ever been in an active state of eruption, and to consider that it consists of porphyry which has been heated in its 'primitive position, penetrated by vapour, and forced up in a mollified state, without having ever flowed as real lithoidal lava.' The form and position of this majestic mountain give it a character of sublimity unequalled even by the loftier heights of the Himalayas. Seen from the coast of the Pacific at 200 miles' distance, it appears as an enormous semi-transparent dome; dim, but the outline too clearly defined against the deep blue of the sky to be mistaken for a cloud.

Cotopaxi, when seen from Quito, is the most beautiful in the whole range, its form being that of a regular and even cone, with a flat summit, covered with a coating of pure white, reflecting the rays of the sun with dazzling splendour. Some idea of its grandeur may be formed if we consider that it is nearly as high above the sea as Vesuvius would be if placed on the top of Mont Blanc. This mountain is in a state of constant activity, the flame rising in the air every night like a colossal beacon. There were five great eruptions in the course of the last century: in 1738 the flame rose to the height of 3000 feet above the summit; in 1743 its sound was heard at Hurda, 200 leagues' distance. On the coast of the Pacific, Humboldt says that its sound is like thunder, or like a continuous discharge of artillery. In 1743 the melted snow descended in torrents, overflowing a distance of five leagues, and the river Tacunga being insufficient to carry off the water, a part of the town was destroyed, and many of the people and cattle carried away. This continued for three days, and the country for more than three leagues round was covered with cinders and scoriæ. In 1768 such quantities of ashes were discharged that the sun was completely hidden, and for many hours the inhabitants were obliged to use lanterns in the streets; and on another occasion a rock of 200 tons weight was ejected to a distance of nine miles.

In January, 1835, the volcano of Coseguina showed signs of activity after twenty-six years' quietude. On the 20th a cloud rose which assumed the form of a plume of white feathers expanding on all sides. This continued for two days, and was then succeeded by intense darkness; a fine white ash fell, and in half an hour it was blacker than night. The fowls went to roost, and men could touch without seeing one another. About noon the following day objects could be seen at ten or twelve yards' distance. This state continued for two days, and for ten or twelve days more the light was partially obscured, during the whole of which time a pure white dust continued to fall, covering the ground in every place in the immediate neighbourhood with ashes, varying in depth from a few inches to upwards of ten feet; and they were carried by the wind upwards of 700 miles, obscuring the sun at Jamaica, and covering the earth there with fine dust.

Other volcanoes are Carguairoso, which in 1698 discharged such a quantity of water, mud, and stones, as covered the ground for nearly 32 square miles, destroying the crops in the fields and many thousands of lives; and Pichincha, which in 1690 discharged a quantity of very fine ashes during twelve days, covering the streets more than two feet thick; and El Altar, or Capururar, which was formerly higher than Chimborazo, but the upper part fell in after an eruption, which also took place with Carguairoso.

Cayambe urcu is the loftiest of the Cordilleras except Chimborazo, to which it bears some resemblance. There are no traditions of its having been in an active state. Pichincha being due west of it, at sunset on a clear evening the shadow of the latter may be observed gradually covering the foreground of Cayambe urcu, and a few seconds before the sun dips in the horizon it ascends with great rapidity, and finally in a moment the whole is lost in darkness. The spectator imagines this is caused by a cloud overshadowing it, and remains gazing in the expectation that the mountain will again emerge; but the briefness of the twilight soon convinces him that he watches in vain, and when he turns his eyes to search for the other mountains they are gone also.

VOLCANIC STORM.

Antisana is an extinct volcano; near its foot is a small village 13,500 feet above the sea, and said to be the highest human habitation in the world.

The phenomena of volcanoes have been the subject of investigation by many eminent scientific observers, but it cannot be said with very successful results. The fact that eruptions take place simultaneously at places very far apart from one another leads to the inference that they originate in the same cause, and that it is deeply seated in the crust of the earth; whether or not that cause consists, as some have supposed, in the electric condition of the earth or its atmosphere, does not seem to be clearly ascertained, though the frequent occurrence of thunder and lightning accompanying an eruption would seem to favour that theory.

That there is some close connexion between the activity of volcanoes and the occurrence of earthquakes seems to be beyond question. Not only are volcanic regions more subject to earthquakes than other points of the world, but the time when they take place seems to show that they arise from the exertion of some agency which labours to find vent in one mode when it cannot in the other. Thus earthquakes are succeeded by eruptions from neighbouring volcanoes, and when volcanoes become less active than usual earthquakes generally succeed. Stromboli had an interval of repose for the first time within the memory of man immediately before the occurrence of the earthquakes of 1783; and when the volcano of Pasto ceased to discharge the dense columns of smoke which it had emitted for three months in 1797, the great earthquake of Riobamba, at a distance of 240 miles, took place, destroying 40,000 persons.

The eruption of Coseguina, already referred to, was accompanied by activity in the volcanoes of Aconcagua and Osorno, at the distance of 2700 miles and 3180 miles, with an earthquake felt over a space of 1000 miles, and was followed within a month by the great earthquake of Valdivia, in which 'seventy villages were destroyed;' in six seconds the town of Concepcion was in ruins; the walls of the Cathedral, four feet in thickness, were thrown down, and the ground was agitated for three days. Half-an-hour afterwards an enormous wave rushed up, thirty feet above high-water mark, overflowing the greater part of the town, and rushed back carrying away with it everything that was movable. When this retired the land was found to be permanently raised, mussels being found adhering to the rocks ten feet above high-water, and acres of coast were covered with dead shell-fish.

These facts, and many more might be cited, are sufficient to prove that there is a connexion between earthquakes and volcanic action, though of what nature is not easy to determine. Whether any such connexion exists between earthquakes and the state of the atmosphere does not appear to be so evident, though many circumstances have given rise to that belief amongst those who live in the countries where they most frequently occur. Even Humboldt, who considers it doubtful, introduces his account of the earthquake which occurred whilst he was

BOLIDE, SEEN IN MINAS GERALS.

at Cumana by a description of the 'very remarkable atmospheric phenomena which were observable.' 'From the 10th of October to the 3rd of November, at nightfall, a reddish vapour arose in the horizon, and covered in a few minutes, with a veil more or less thick, the azure vault of the sky.' 'In the night between the 3rd and 4th of November (the day when the earthquake took place) the reddish vapour became so thick that I could not distinguish the situation of the moon except by a beautiful halo of 20° diameter.' And after the account of the earthquake, he proceeds to describe an extraordinary shower of meteors

BOLIDE.

which took place a few days afterwards, when 'there was not in the firmament a space equal to three diameters of the moon which was not filled every instant with bolides and falling stars.' He says elsewhere, that he 'never beheld meteors so multiplied as in the vicinity of the volcanoes of the province of Quito, and in that part of the Pacific Ocean which bathes the volcanic coasts of Guatemala.' Meteors are also frequent in Chili, another volcanic region.

Bolides are not unfrequent in non-volcanic districts. M. Liais saw one, while journeying in Minas Geraes on a bright moonlight night, which suddenly separated into two parts, the larger portion continuing its progress in the same direction, and the other taking a course at a considerable angle.

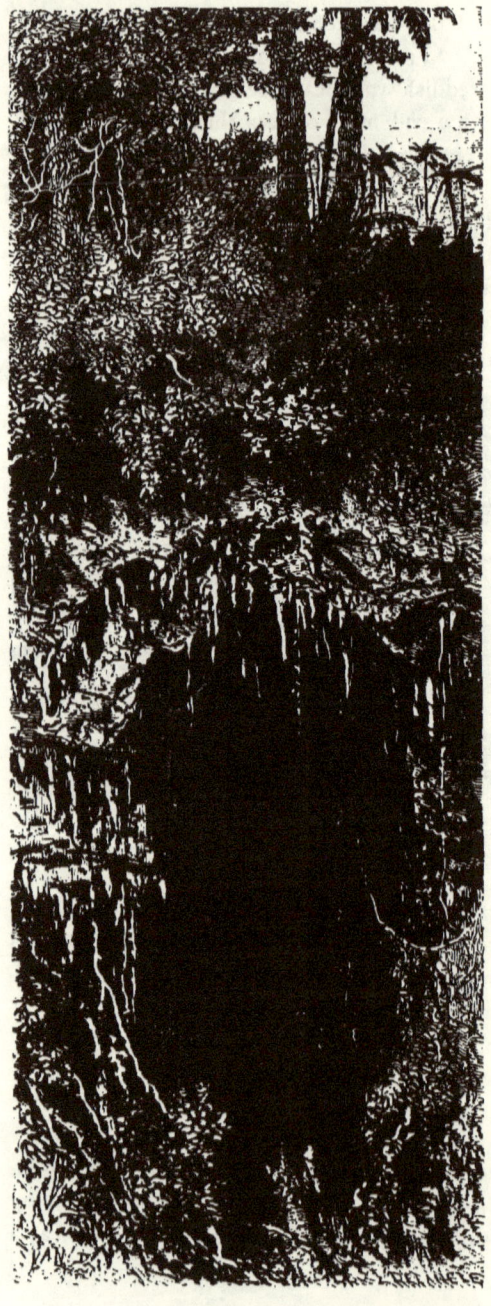

CAVERN.

The valley of Rio das Velhas, in the province of Minas Geraës, is probably the most abundant in caves of stalactites of any part of the world. It contains more than 1000 caverns, the greater part of which have never been examined —some only having been worked for the sake of the salt-petre in which they abound. Many bones have been found in them, of species of which many are extinct, but nearly all belonging to genera still existing in America. 'It is frequently necessary,' says the author from whom we quote, ' to cut a way through the forest of lianas and all kinds of shrubs before they can be entered; and when, after having travelled for hours through paths thus opened, we arrived at an enormous mass of chalk, it was necessary to devise some means of attaining the almost inaccessible entrance of the cave. But we were well repaid for all our

STALACTITE CAVE NEAR JAGUARA.

labour when we succeeded in gaining an entrance. Sometimes, after passing through a narrow passage, we entered a vast hall decorated with picturesque garlands formed of stalactites, and surrounded by crystalline walls, which reflected the light of our torches in thousands of rays. A series of passages—sometimes wide and lofty, sometimes narrow, low, and winding—led us from one to another of these halls, some of which rivalled our cathedrals in size. Some of these caverns are connected with passages, which we were told were more than a mile long, having separate entrances. At the entrance of one of the most beautiful of these halls, called the *cave of the great rains*, there was an enormous bell formed by stalactites, which gave a sound when it was struck. We saw few traces of animals; though abundant in the neighbourhood, they do not appear to frequent the caves.'

But though the quadrupeds do not inhabit these caves, there are some which are used as dwellings by immense numbers of birds. Humboldt describes one

ENTRANCE OF CAVE.

of these which he visited near the Missions of Caripé: it is situated about three leagues to the south-west of the convent; and the birds, which are killed for the sake of their fat, are so numerous that it is called by the Indians by a name which signifies 'a mine of fat.' The path was sometimes by the side of a stream which rises in the cave, and sometimes through the stream: they did not perceive the entrance till they were within four hundred paces of it. The mouth of the cave is an arch eighty feet broad, and upwards of seventy in height. The rock above is covered with gigantic trees, and even within the cave the vegetation continues for thirty or forty paces from the entrance. The cave continuing in the same direction, they did not find it necessary to light their torches till they had proceeded about four hundred and thirty feet.

The guacharo, by which this cave is inhabited, is a bird about the size of an ordinary fowl, the wings when open spreading to three feet and a half in width. It is of a dark bluish-grey colour, mixed with small streaks and spots of black; the head, wings, and tail are marked with large white spots in the shape of a heart, edged with black. It quits the cave at night, and is almost the only nocturnal bird known that feeds on fruits. The Indians showed the travellers the nests by fixing a torch to the end of a long pole. They were fifty or sixty feet above their heads, in funnel-shaped holes, with which the roof of the grotto was pierced. The shrill and piercing cries of the birds were repeated by the echoes of the vaults, increasing as they advanced and the birds were disturbed.

The Indians enter the Cueva del Guacharo once a-year near midsummer, armed with poles, with which they destroy the greater part of the nests; the old birds, as if to defend their brood, hovering over the heads of the Indians uttering loud cries. The young birds which fall to the ground are opened on the spot, and are found loaded with fat, forming a kind of cushion between their legs. This fat is melted in pots of clay, and is known by the name of butter, or oil of the gaucharo. It is half-liquid, transparent, without smell, and so pure that it may be kept above a year without becoming rancid.

The natives are very superstitious, and the travellers had great difficulty in persuading them to pass beyond the portion of the grotto which they annually visit, although supported by the whole authority of 'los padres.' They believe that the souls of their ancestors inhabit the recesses of the cave; 'to go and join the guacharos' is with them a phrase signifying to rejoin their fathers—to die.